FEATURE BASED PRODUCT LIFE-CYCLE MODELLING

IFIP - The International Federation for Information Processing

IFIP was founded in 1960 under the auspices of UNESCO, following the First World Computer Congress held in Paris the previous year. An umbrella organization for societies working in information processing, IFIP's aim is two-fold: to support information processing within its member countries and to encourage technology transfer to developing nations. As its mission statement clearly states,

IFIP's mission is to be the leading, truly international, apolitical organization which encourages and assists in the development, exploitation and application of information technology for the benefit of all people.

IFIP is a non-profitmaking organization, run almost solely by 2500 volunteers. It operates through a number of technical committees, which organize events and publications. IFIP's events range from an international congress to local seminars, but the most important are:

- The IFIP World Computer Congress, held every second year;
- open conferences;
- working conferences.

The flagship event is the IFIP World Computer Congress, at which both invited and contributed papers are presented. Contributed papers are rigorously refereed and the rejection rate is high.

As with the Congress, participation in the open conferences is open to all and papers may be invited or submitted. Again, submitted papers are stringently refereed.

The working conferences are structured differently. They are usually run by a working group and attendance is small and by invitation only. Their purpose is to create an atmosphere conducive to innovation and development. Refereeing is less rigorous and papers are subjected to extensive group discussion.

Publications arising from IFIP events vary. The papers presented at the IFIP World Computer Congress and at open conferences are published as conference proceedings, while the results of the working conferences are often published as collections of selected and edited papers.

Any national society whose primary activity is in information may apply to become a full member of IFIP, although full membership is restricted to one society per country. Full members are entitled to vote at the annual General Assembly, National societies preferring a less committed involvement may apply for associate or corresponding membership. Associate members enjoy the same benefits as full members, but without voting rights. Corresponding members are not represented in IFIP bodies. Affiliated membership is open to non-national societies, and individual and honorary membership schemes are also offered.

FEATURE BASED PRODUCT LIFE-CYCLE MODELLING

IFIP TC5 / WG5.2 & WG5.3 Conference on Feature Modelling and Advanced Design-for-the-Life-Cycle Systems (FEATS 2001)
June 12–14, 2001, Valenciennes, France

Edited by

René Soenen
University Claude Bernard of Lyon
France

Gustav J. Olling
DaimlerChrysler Corporation
USA

SPRINGER SCIENCE+BUSINESS MEDIA, LLC

Library of Congress Cataloging-in-Publication Data

A C.I.P. Catalogue record for this book is available from the Library of Congress.

Feature Based Product Life-Cycle Modelling
Edited by René Soenen and Gustav J. Olling
ISBN 978-1-4757-5213-7 ISBN 978-0-387-35637-2 (eBook)
DOI 10.1007/978-0-387-35637-2

Printed on acid-free paper.

Contents

Contributors

van den BERG J., Delft University of Technology, The Netherlands
BORG J.C., University of Malta, Malta
BRONSVOORT W.F., Delft University of Technology, The Netherlands
BRUNETTI G., Fraunhofer-Institut (IGD), Darmstadt, Germany
CANCIGLIERI J.O., University of Parana, Curitiba, Brazil
CHOI B.K., KAIST, Seoul, Korea
CUGINI U., Politecnico di Milano, Italy
FRANK D., DaimlerChrysler AG, Ulm, Germany
GIANNINI F., CNR, Genova, Italy
GIUDICE F., University of Catania, Italy
HAASIS S., DaimlerChrysler AG, Ulm, Germany
HOEK F.M., Delft University of Technology, The Netherlands
KIM B.H., KIIT, Seoul, Korea
KIM D.H., Wooyoung Co., Seoul, Korea
KIMURA F., University of Tokyo, Japan
KO K., KAIST, Seoul, Korea
LA ROSA G., University of Catania, Italy
MANDORLI F., University of Ancona, Italy
MATSUDA M., Kanagawa Institute of Technology, Japan
NOORT A., Delft University of Technology, The Netherlands
RISITANO A., University of Catania, Italy
RIZZI C., University of Parma, Italy
ROMMEL B., DaimlerChrysler AG, Ulm, Germany
SUSCA L., University of Parma, Italy
SUZUKI H., Arthur D. Little, Japan
WEYRICH M., DaimlerChrysler AG, Stuttgart, Germany
YOUNG R.I.M., Loughborough University, United Kingdom

Acknowledgements

IFIP	International Federation for Information Processing TC5: Computer Applications in Technology WG 5.2: Computer Aided Design WG 5.3: Computer Aided Manufacturing
CIRP	International Institution for Production Engineering Research
IEEE	Institute of Electrical and Electronics Engineers SMC Society: Systems, Man and Cybernetics
IFAC	International Federation of Automatic Control TC-MIT: Advanced Manufacturing Technology TC-MIM: Manufacturing Modelling, Management and Control TC-SMC: Computer Aided Control Systems Design
EC-IST	European Commission IST: Information Society Technologies
MR	Ministère de la Recherche
CNRS	Centre National de la Recherche Scientifique
CR-NPDC	Conseil Régional Nord Pas de Calais
GRAISYHM	Groupement de Recherche en Automatisation Intégrée et Systèmes Hommes/Machines
UVHC	Université de Valenciennes et du Hainaut-Cambrésis
LAMIH	Laboratoire d'Automatique de Mécanique et d'Informatique Industrielles et Humaines (UMR CNRS-UVHC n° 8530)

Preface

The customers of the twenty-first century are definitely demanding products that bring together innovation, quality, immediate availability, safety, low cost and even service. In that context, the design activity in charge of elaborating, evaluating and validating candidate products is faced with the complex problem of optimising upon the product's life cycle many conflicting parameters relating to heterogeneous points of view (functions, technology, machining, assembly, time to market, legal aspects, use, wear, global cost, environmental impact, etc.).

Since the maturity of Computer Aided Design systems, feature modelling has been considered as a powerful means to characterise the product from various points of view.

This book gathers a selection of contributions to the International IFIP Conference on Feature Modelling in Advanced Design-for-the-Life-Cycle Systems (FEATS 2001), which convened in Valenciennes, France in June 2001, providing a comprehensive state of the art and future trends about the use of feature technology in advanced design systems.

Dr D. DENEUX
Chairman of FEATS'2001

Exploiting Integrated 'Product' & 'Life-Phase' Features

J.C. BORG[1] and F. GIANNINI[2]
[1]*Department of Manufacturing Engineering, University of Malta, Malta,*
[2] *Istituto di Matematica Applicata e Tecnologie Informatiche, CNR, Genova, Italy*

Abstract: Increasing competition requires more and more effort in delivering new products with better prices, good quality and environment savings. In this scenario, since most of the cost and product characteristics are dependent on commitments taken at the early design stages, designers require tools supporting them at considering the *consequences* of their decisions on the whole product life-cycle, starting from the conceptual design stage to the disposal phase. The adoption of such tools may enforce the exploration of different alternatives thus increasing the possibility of identifying the most convenient and innovative solution. Form features have been recognized as shape-oriented elements for associating geometry with engineering meaning, thus helping in reasoning on products in functional terms. However as argued in this paper, information on shape alone is insufficient for meaningful evaluation and forecast of life-cycle product consequences. The paper presents an approach for considering 'life-cycle consequences' *during* the design decision process, by taking into account both artefact features and the characteristics of the involved life-cycle systems.

Key words: Product features, Life-phase features, Life-cycle oriented design, Reuse.

1. INTRODUCTION

Design decisions can result in *unintended consequences* that have a *propagation effect* across *multiple* product life-phases such as manufacturing, use and disposal. Strong market competition and the increasing trend to environmental care are demanding major attention during product development for considering all the life-cycle issues. Thus, rather than a *narrow* 'Design for X' (DFX) approach, designers need to adopt a 'Life-oriented Design' (LOD) approach. Such a LOD approach treats *Xs*, like life-phases (e.g. disposal) and performance measures (e.g. cost) in a *multiple* and *integrated* way during synthesis. Since it has been demonstrated that most of the product characteristics and costs depend on decisions taken at the early design stages (Andreasen et al. 1997), DFX-type knowledge should be available and used as from the conceptual design stage when the solution is still evolving. However, current manual and computer based design approaches, do not adequately support designers in handling the phenomena of propagation effects during synthesis, by *foreseeing* and exploring life-cycle consequences (LCCs) co-evolving with their *synthesis decisions* (Borg 1999). As argued in (Tomiyama 1998), for generating more added values, CAD tools must include the life-cycle knowledge. To address this situation for mechanical component life design, this paper underlines the need of tools helping designers in revealing and evaluating the consequences of their decisions simultaneously in the different environments the product will be used or manipulated, with the aim of supporting them in exploring different alternatives in order to find the best solution. In this paper, the necessity of introducing life-cycle knowledge linked to the characteristics describing products, such as form features and material is discussed. To this aim, the paper presents the concept of a life-oriented design tool employing both *artefact features* and *life-phase features*.

The paper is organized as follows. Section 2 illustrates advantages and limitations of the current use of feature technology emphasizing the need of managing additional information beyond shape. Section 3 presents a model of the decision making process and how LCCs are generated during life-oriented design, based on the consideration of the intended and unintended consequences of the chosen solution elements. Section 4 provides an overview of the developed prototype CAD system (Borg 1999) supporting designers in considering the life-cycle consequences (LCCs) co-evolving with commitments made during early design. In section 5, current limitations of the prototype and future improvements are discussed, whilst conclusions are presented in section 6.

2. CURRENT FEATURE-BASED TECHNOLOGY

The demand of supporting and utilizing engineering knowledge beyond simple geometry has resulted in the introduction of the feature concept (Cunnigan 1987, Shah 1991, Shah 1995). A feature can be seen as a reusable element, having a predictable shape and behaviour that can be associated to some engineering function or operation.

It is possible to say that, feature-based solution models provide a high level description of parts in terms of significant elements. Thus, focusing on the most important context-dependent characteristics and leaving out the basic geometric details, they support effective reasoning on the parts

Even if no argument about the importance of features exists, their real power is far from exploited. Current commercial CAD systems support the use of form features during the detail design modelling activity in terms of predefined parametric macros, corresponding to the most recurrent shapes, but no additional engineering information or consistency check on their semantic is provided. As an example consider the basic feature 'blind hole', no checks are performed to verify if during the design solution evolution it is transformed to a through hole due to some unintended interactions with added product details. This transformation can change completely the functional characteristic of the feature, e.g. from a blind cavity with only one access to a passage through which an inserted element can move through. Analogously, the attempts to automatically integrate CAD with production and analysis are still limited (Part™, FeatureWorks™, CAMWorks™) and mainly at research level (Henderson 1984, Regli 1995, Falcidieno 1989, Pinnilla 1989, Vandenbrande 1990). This is mainly due to the fact that such systems mainly focus on the geometric description of the features, and the use of fundamental information related to other solution characteristics, such as material and the available technologies, is still limited. *Figure 1* shows that even considering the same context, *e.g. manufacturing*, a feature-based description of a mechanical part can be deeply different as this also depends on the product material and the adopted technology. Thus, the same design object can be viewed in terms of two volumes to be removed if it is made of mild steel and has to be milled. Alternatively, it can be viewed to consist of a protruding rib if it requires a mould specification in case it is made from a plastic material.

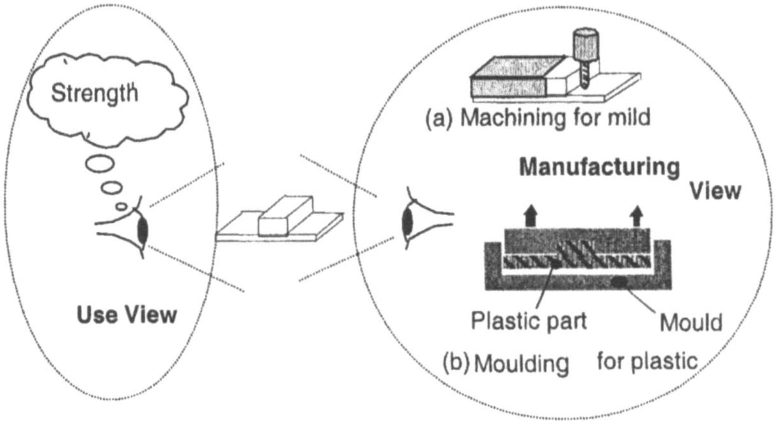

Figure 1. An example of describing the same part from different viewpoints

With the design process shifting from sequential to concurrent, the simultaneous availability of the two approaches has been required to make available the different context dependent representations to the involved experts. Thus, some activities have been focused on integrating the two approaches (Sreevalson 1990, De Martino 1994, De Kraker 1995).

At present features are mainly used during the solution synthesis activity of the detailed design stage, when details are being specified, or during the evaluation phase, when all the design commitments are almost fixed and all the fundamental decisions have been already taken. Thus changes are difficult to make and are time consuming. This is why it makes sense to consider life-cycle consequences as from early design. Some work in this area exploiting the concept of features is going on as reported (Borg 1999). However, up to now, no real support for the evolution of the design solution from the conceptual stage to the detail one is provided. Features appear to be among the key elements on which to base such an evolution (Krause 1996).

3. LIFE-CYCLE ORIENTED DESIGN DECISION MAKING

Artefact life-oriented design (LOD) implies the consideration of all the transformation activities and phenomena an artefact has to encounter right from its realization use until its disposal. To do so, it is necessary to consider the relationships among the elements describing the artefact and the involved phenomena (Otto 1998, Vajna 1997). In fact as argued in (Borg 1999), the *interaction* of an artefact with either *natural* (e.g ocean) or *artificial* (e.g. milling machine) life-phase systems (*Figure* 2) can give rise to a number of *unintended life-cycle consequences (LCCs).* If the artefact had to interact with a *different* set and sequence of life-phase systems, different LCCs would result.

A LOD approach therefore requires that designers *foresee* what life-phase systems will be met during the life of an artefact and that they also foresee the outcome (consequences) of such *interactions* during design. Due to the chronological order of life-phases, designers do not generally *acquire* experiential knowledge concerning LCCs resulting from artefacts interacting with different life-phase systems. Rather, such experiential knowledge is *acquired* and *distributed* amongst various human actors (e.g. assembly operators, machining operators, users, servicing engineers) who are *distributed*, internally and externally to an organization. Hence, due to a lack of LCC knowledge, design decision-making takes a narrow and segmented view, this affecting the quality of the solution with respect to artefact life issues. Therefore a LOD approach requires a vast amount of distributed knowledge to be acquired, readily available and easy to access in order for it to be explicitly utilized during *synthesis*. Understanding the phenomena of how LCCs are generated and propagated during *synthesis decision-making* is thus essential for the development of CAD tools supporting LOD.

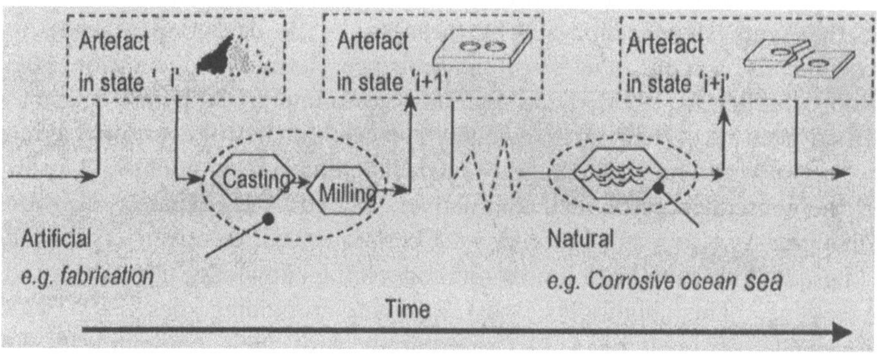

Figure 2. Interactions between artefact and life-phase systems

3.1 'Artefact Life' design decision-making

A mechanical artefact can be considered as a decomposable system (Hubka. 1988) consisting of a number of *re-usable* elements, e.g. sub-assemblies, composed in turn of simpler elements such as form features, assembly features, material and surface textures. All these elements, termed in this paper as *Product Design Elements (PDEs)*, are related to each other with *'part of'* relationships. Similarly, the life-phases (e.g. *design, realization, use, disposal)* forming a mechanical artefact's life, involve the *re-use* of technical systems (e.g. manufacturing or maintenance systems) that realize the relevant transformation effects of that phase. Models of these systems (e.g. milling machine) can be also decomposed into sub-systems (e.g. a work piece holding device), these termed *Life-Cycle Phase Elements (LCPE).* Such a decomposition supports life-phase system compositional modelling, where *LCPEs* are *related* to *'part of'* relationships.

The design process can be seen as a decision intensive activity, which involves the simultaneous specification of the artefact (i.e. using relevant PDEs) and the technical systems involved in the different life-phases (i.e. using relevant LCPEs). Regarding the artefact, irrespective of the specific *design stage* (e.g. conceptual vs. detail) or *synthesis viewpoint* (e.g. functional or constructional), the decision maker has to make a choice between a numbers of alternatives related to the domain (March 1994). In the case of components, the alternatives concern a number of manipulable characteristics (Tjalve 1979), which include reusable PDEs. Similarly, the synthesis of life-phase compositional models involves the reuse of LCPEs. In *Figure 3*, some examples of typical alternatives in mechanical design are shown.

The alternative selected, termed here the decision commitment, is a result of the synthesis decision-making process. A decision is made to *intentionally* achieve a *desired* consequence, termed the decision goal (Roozenburg et al. 1995). Thus, alternatives are interpreted by the decision-maker, in terms of their *expected* consequences. Studies of decision making in the *real* world however suggest that decision-makers do not always know *all* the consequences of their alternatives (March 1994). That is, decision commitments also result in *unintended* consequences. Designers, as human beings, are known to have limitations concerning knowledge possession and processing. This highlights why *explicitly* providing designers with knowledge of unintended LCCs resulting with their commitments, is beneficial to supporting them in selecting 'life-oriented alternatives'. These commitments, made during synthesis to an evolving solution, are termed *synthesis decision commitments* (Borg 1999).

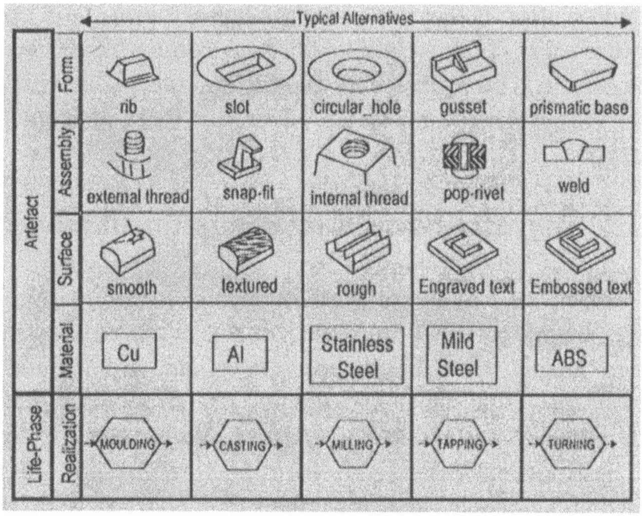

Figure 3. Some typical alternatives in mechanical artefact life design

Unlike other commitments (e.g. about the design process), synthesis decision commitments are directly reflected in the evolving artefact's solution model (*Figure 4*).

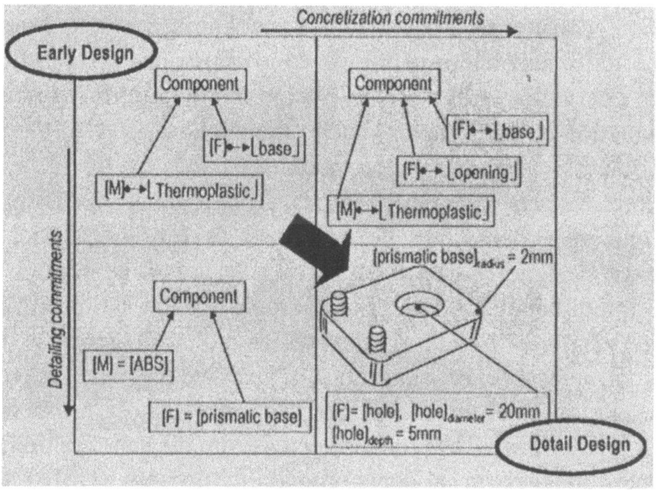

Figure 4. Artefact model evolution with synthesis decision commitments

Through the *theory of domains* (Andreasen 1992), it can be said that these commitments result in a more concrete solution if they are *concretisation commitments* (e.g. committing an opening form feature) or a more detailed solution if they are *detailing commitments* (e.g. committing a value for a diameter). Since synthesis decision commitments become part of the artefact solution, then designers should be concerned with *unintended LCCs* arising from such commitments. Modelling these phenomena is useful for this research.

3.2 A synthesis decision commitment model

Figure 4 discloses a model of synthesis decision-making upon which the approach framework to LOD developed in this research is based. Due to consequences arising from the *interaction* of an artefact and life-phase systems, this paper argues that life-oriented design requires that designers, concurrently generate and model the *artefact solution* and *life-phase system solutions*. These models collectively form an *'artefact life' model*. During such concurrent synthesis, decisions therefore concern both the *artefact model* and the different *life-phase models*. Without concurrent synthesis, consequences arising from the interaction of specific artefact solutions and specific life-phase systems would be difficult to cater for. This need can be explained via the example in *Figure 1*. Depending on the material defined for the part, different processing technologies are required during the manufacturing phase. This introduces *different* consequences that can in their turn affect the product specification. For instance, to obtain a feasible part solution, constraints related to the machining have to be considered, such as the minimum width of the rib to avoid cracks/spring effects during the milling operation. Alternatively, shape modifications may be necessary (e.g. introduction of draft angles) due to the sticking effect if the plastic part is to be efficiently ejected from the mould.

Therefore, for this research purposes, from a generic point of view, this decision making process can be described as follows. Basically, during synthesis, the designer *generates* a set {O} of possible solution options to the sub-problem being tackled. For example for the sub-problem *how can two parts be joined together?* the set of solution options {O}may be {fasteners or adhesive or snap-fits}. Thus, in the case of routine and innovative design, the set of options {O} consists of reusable PDEs or LCPEs. As an *alternative* has to be selected from the set of options available, the designer engages in a decision-making process (DMP) to make a selection. During the DMP, the designer considers *intentions*, *preferences* and known *circumstances*. Following these considerations, the designer selects an alternative by making a synthesis decision commitment $[d]_E\{O\}$ to the evolving solution model (*Figure 5*). The synthesis decision commitment model in *Figure 5* is also applicable to life-phase system synthesis and exploration. In the latter case, the solution model being generated and explored is that of a life-phase system and the decision-making process concerns feasible, alternative LCPEs (e.g. alternative fabrication systems). Life-phase synthesis decisions concern for instance the selection of technical systems (e.g. milling) and system parameters (e.g. a feed-rate value). In order to develop a tool supporting designers in foreseeing LCCs evolving with their commitments, it is therefore important to understand how LCCs are generated.

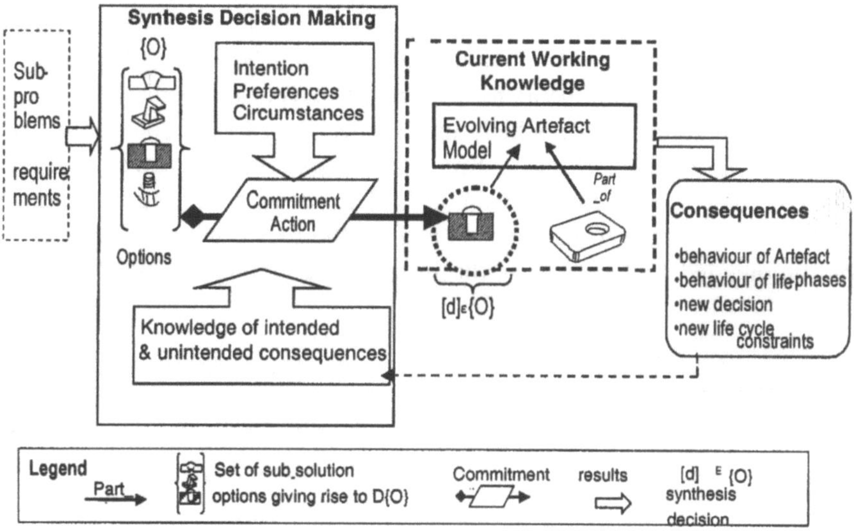

Figure 5. Synthesis decision commitment model

3.3 Life-cycle consequences generation

Based on the foregoing arguments, it can be noted that LCCs result from two fundamentally different conditions:

i. individual, *non-interacting* synthesis decision commitments, resulting in what are termed, non-interacting LCCs, i.e. **LCC_{ni}**. For example, an LCC_{ni} resulting from the commitment of a pop-rivet assembly feature to bind two separate parts is that 'dis-assembly of the parts in the disposal phase is slow';

ii. multiple and *interacting* synthesis decision commitments, resulting in what are termed, interacting LCCs i.e. **LCC_i**. For example, an LCC_i resulting from the commitment of a hole *and* the commitment of sheet-metal as a part's material is that 'a suitable punch needs to be used during fabrication in the realization phase'.

Moreover, the knowledge about LCCs is *implicitly* co-evolving with *concretisation* and *detailing* synthesis decision commitments being made to the artefact life solution. In fact, it is possible to say that whilst the set of possible solutions reduces with the *detailing* and *concretisation* of the commitments made, in the meantime, the determination of the LCCs becomes more certain and covering more situations. Some LCC can be deduced at the conceptual stage, e.g. for a plastic part, a mould has to be defined, while most of them can only be deduced precisely during the detailing stage. This concerns mainly the case of consequences related to dimensional and positional values that give rise to specific feature interactions both within the artefact and among the artefact and life-phase system. Examples of such interactions are shown in *Figure 6*. In *Figure 6*(a), the concretisation commitments for the two PDEs (pocket features) give rise to a thin separating wall, which under specific LCPEs conditions can produce an unintended crack. Similarly, *Figure 6*(b) demonstrates that concretisation commitments of two holes (PDEs) to be generated via a punch press (an LCPE) have to be checked against the LCPE features in order to avoid unintended LCCs, such as the interference among the punching tool holders.

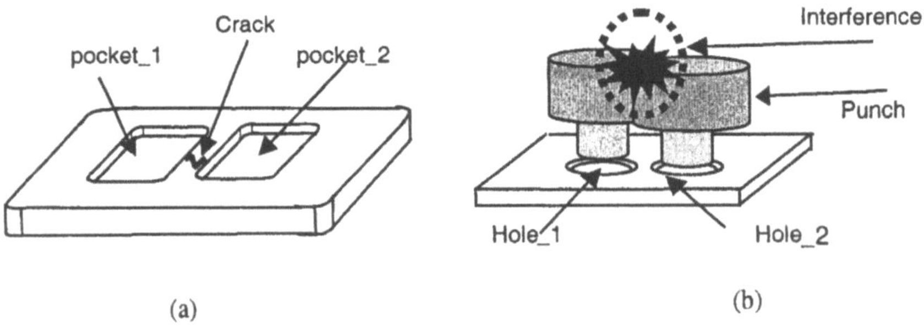

Figure 6. Examples of interactions giving rise to unintended LCCs: (a) interaction of PDE-PDEs features (b) interaction of PDE - LCPE features

4. TOWARDS FEATURE-BASED COMPONENT 'LIFE ORIENTED' DESIGN TOOLS

The phenomenon of co-evolving LCCs therefore discloses that *during* artefact life design, knowledge is both *used* and *generated.* If the co-evolving *LCC knowledge* is explicitly revealed and utilized *during* synthesis decision-making, it provides *additional* knowledge that can be employed by designers for comparing alternative PDEs / LCPEs in terms of their resultant LCCs. If made aware of unintended consequences, a designer may therefore retract a synthesis commitment to explore other alternatives. By revealing solution-specific problematic LCCs, designers can be motivated to explore alternative commitments, thereby being guided in the generation of *life-oriented* design solutions. In order to *timely* support the designer's thinking process, designers need principles disclosing how to describe evolving artefact life solutions from which co-evolving LCCs can be inferred. Further, to adopt this approach, there is a need to describe *what* needs to be captured and modelled to generate a suitable LCC knowledge base. Therefore, based on the decision commitment model (*Figure 5*),'Knowledge *of life-cycle Consequences' (KC)* approach framework has been developed, this composed of an *artefact life modelling frame*, a *knowledge-modelling frame* and an *operational frame.*

4.1 Artefact life modelling frame

In this approach, an 'artefact life' solution is described in terms of an 'artefact model' and related 'life-phase system models', these models described via reusable PDE and LCPE models respectively. Basically, a PDE model describes reusable 'artefact system' *design elements.* For instance, for component synthesis from a constructional viewpoint, typical PDE models are those of form *features* or of materials. PDEs that have been found relevant for designing mechanical components from a constructional viewpoint are *form features, assembly features, material, surface finish, dimensions* and *tolerances*, these detailed in (Borg 1999). For example, this research identified a number of characteristics that need to be used for describing assembly feature PDEs as they give rise to LCC_{ni} – *assembly feature joint strength, joint dynamism, integrity, joint permanence* and *assembly repetitivity.*

For 'life-phase system modelling', designers are concerned with manipulating reusable life-phase system elements. As with artefacts, the synthesis of a life-phase system models can take place from different viewpoints. Thus a life-phase model $\langle Phase\rangle_i$ is considered in this paper to be composed of a set of *transformation systems* (from a constructional viewpoint) delivering *transformation process [P]* effects. For instance, a fabrication system (see *Figure* 2) can be composed into a casting process followed by a milling process. LCPEs are thus reusable units (e.g. milling machine) and sub-elements (e.g. workpiece fixture) that make up a transformation system delivering the *processing* effects that transform the artefact from one state to another during this interaction. The generic LCPE model developed in this research is based on Hubka & Eder's (Hubka et al. 1988) system model of a transformation process. The input *operand* Od^1 receives effects based on a *process technology* $[P]_t$ (e.g. erosion by sparks) to be transformed into *operand state* Od^2. These transformation effects are delivered through *human beings* (e.g. machining operators) and *technical systems [TS]*, the latter basically decomposable into *executing systems (e.g. tooling)* and *control systems.* As reflected by non-exhaustive examples in Table 1 derived from (Borg 1999), this model applies to transformation processes encountered in different artefact life-phases.

As explained in (Borg 1999), LCPE characteristics that have been considered sufficient for describing LCPE models supporting the inference of LCCs from the interaction of a conceptual mechanical component solution and a conceptual life-phase system solution are: *process technology, technological properties, process parameters, process parameter values, process minimum economic quantity, process technical (physical) system,* and *technical system parameters.* As the current focus has been on supporting early design, certain characteristics have been omitted from this LCPE model e.g. the material of the elements making up a [TS], the human operator (e.g. a novice *versus* an experienced machining operator), the assembly features used in a [TS], the control system of a [TS] and the environment (e.g. humid) in which the [TS] operates.

Table 1 Some life-phase transformation processes & their typical characteristics

Life-phase	*[P]*	*[TS]*	$[P]_t$	Od^1	Od^2
Realization (fabrication)	*Spark Erosion*	*Spark eroder*	*Material erosion by electric sparks*	*Initial component form*	*Different components form*
Realization (fabrication)	*Sand casting*	*Sand casting mould*	*Material solidification*	*Raw material*	*Cast component*
Realization (assembly)	*Arc welding*	*Welding machine*	*Electric power material fusion*	*Separate components*	*Bonded components*
Use (maintenance)	*Grit blasting*	*Grid blasting device*	*Chip formation by material abrasion*	*Dirty artefact surface*	*Clean artefact surface*
Disposal	*Magn. separation*	*Separating machine*	*Magnetic attraction*	*Mixed components*	*Separated components*

4.2 LCC knowledge modelling frame

This frame concerns the formalization and definition of the following:

a) ***a life synthesis element library:*** this consists of a structured library of various PDE and LCPE models reused within the design and life of an artefact's domain. Elements in the library can be generic (e.g. a hole form feature) or domain specific (e.g. mould core-pin). This library extends the designer's mental knowledge base with reusable synthesis elements used for generating 'artefact life' solutions.

b) ***relationships*** between PDEs/LCPEs and associated LCCs describing :

   i. *LCC inference knowledge*: this consists of descriptions *logically* relating artefact domain synthesis elements to LCC_{ni} and/or LCC_is. Based on the phenomena model, in the case of LCC_is, interacting relationships can be between different PDEs, different LCPEs or between PDEs and LCPEs;

   ii. *LCC action knowledge* describing:

      - *concurrent synthesis patterns:* these correspond to default commitments(e.g. add core-pin to mould tool) associated with certain LCCs (e.g. core-pin required *Figure* 7–b) that allow the automatic evolution of a specific model (e.g. moulding machine model
      - *mappings between LCCs and performance measures* – a LCC (e.g. sink mark) can cause a change to a *performance measure* (e.g. cost), of a technical process [P] (e.g. moulding), forming part of a phase (e.g. realization). Changes in performance measure values are modelled *relative* to values resulting from other possibilities, e.g. on a range –10 to +10.
      - *explanations of specific LCC and guidance to their avoidance/relaxation*: having inferred a LCC (e.g. a weld line defect), designers need explanation of what the LCC means and what it effects, together with guidance to its avoidance/relaxation. Through the LCC phenomena model, knowledge of *which* commitment(s) give rise to a detected LCC can be made explicit, thus guiding designers to the list of alternative commitments that can be explored.

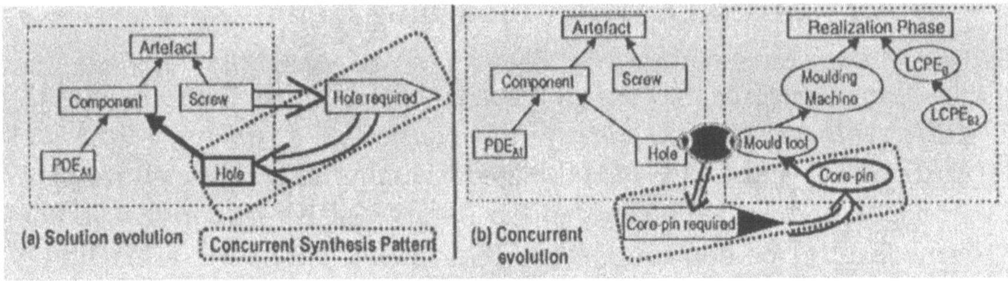

Figure 7. Typical concurrent synthesis patterns

4.3 Operational Frame

The operational frame provides a LOD *working environment* and *mode of operation.* Basically, through this frame and the other frames in the 'KC' approach framework, when a design sub-problem is encountered (step 1), the designer can interact with a synthesis elements library to search (step 2) for a set of suitable elements (step 3). Based on the designer's intentions, preferences and circumstances, elements are committed (step 4) to evolve the 'artefact life' model. This evolving model is monitored (step 5) by *LCC inference knowledge* that detects (step 6) any co-evolving LCCs. Relevant *LCC action knowledge* infers actions that need to be carried out, such as changes in performance measures of appropriate life-phase metrics to allow designers to monitor the artefact life behaviour. Collectively, this inferred co-evolving LCC knowledge is utilized (step 7) for exploring the avoidance/relaxation of the detected LCCs.

4.4 Prototype CAD System

The 'KC' approach framework described has been implemented as a prototype CAD system, named FORESEE. As implemented, it allows designers to generate and describe (i) an early mechanical component compositional model and (ii) a number of early life-phase system compositional models. It reveals a list of LCCs associated with the current solution state and an associated set of *multi-X* relative performance measure values to proactively guide the designer to monitor the performance of the evolving artefact life solution. FORESEE also provides explanations of detected LCCs and guidance to their avoidance. Further, a history of the decision commitments made and their associated impact on performance measures is provided. The system details and a typical design scenario of its use are presented in (Borg et al. 1998). In *Figure 8*, some snapshots of the developed prototype are shown.

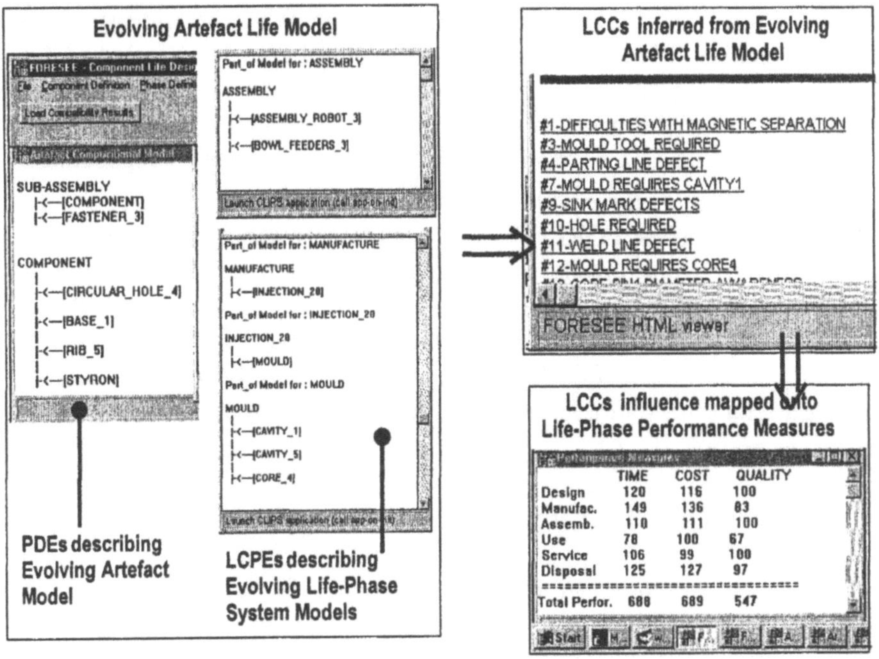

Figure 8. Sample screen shots of the prototype tool FORESEE

5. PROTOTYPE TOOL EVALUATION AND FUTURE EXTENSIONS

The implementation demonstrated that the knowledge intensive 'KC' approach framework can be realized using existing computer technology. Via the FORESEE prototype and an empirical evaluation, strengths and limitations of this approach have been established (Borg 1999). Since 91% of the evaluators considered the integrated 'product' and 'life-phase' feature-based LOD approach to be useful in practice, further research is merited to address the revealed weaknesses such as:

(i) Feature-based LOD approach improvements

The current approach provides the user with general LCC which depend on specific PDEs and LPEs assumptions. What would be interesting is the possibility of evaluating a solution generated in terms of PDEs associated with alternative LCPEs e.g. to evaluate the LCCs that result from changing the producer, and consequently the system (i.e. LCPEs), for that specific component. Also required is support for the evolution of the component solution in terms of PDEs and in particular feature instantiation since, as argued in section 3.3, their interactions strongly effects the detected LCCs. This means that algorithms for the recognition of feature interactions should be applied, and if the case, to re-organize the description to check if the previously obtained LCCs are still valid, or if new ones have been generated.

(iv) FORESEE Prototype System Improvements

For practical purposes, FORESEE also needs improvements. For example, the knowledge management facility needs to be extended to support distributed life-actors to *input* any new LCC knowledge they encounter in the *right modelling format* and in *validating* the input knowledge. Work is also required to enable FORESEE to simultaneously handle and display *multiple* artefact life solutions for easier comparison purposes.

6. CONCLUSION

This paper presented an integrated 'product feature' and 'life-phase feature' *based* approach for supporting 'component life design'. Its implementation, the FORESEE prototype, discloses that the approach contributes a step towards the development of *Knowledge Intensive CAD* (KICAD) tools aimed at pro-actively supporting designers in generating 'life-oriented' artefact solutions. As established, the approach concept:

- reflects that a set of *reusable* synthesis elements i.e. PDEs to describe 'artefact models' and LCPEs for describing 'life-phase system models' are required when developing *life-oriented* KICAD tools. In this sense, PDEs and LCPEs extend *feature based 'artefact' design* to *early 'artefact life' solution design*;
- demonstrates that KICAD tools provide a suitable means to retain, process and explicitly *reuse* captured LCC knowledge for guiding designers in LOD;
- reflects that the general system requirements on which the FORESEE prototype is based, contribute to the development of LOD tools;

As discussed, further work is however required to this knowledge intensive feature-based approach to LOD if it is to practically support the design process as from its early conceptual stage to the detailed stage. As a first step, this requires the specification of models able to represent reusable elements such as geometry and material in a vague way, together with the development of algorithms for feature interaction recognition and mapping from such incomplete solution descriptions.

REFERENCES

Andreasen M.M. (1992) Designing on a 'DesignersWorkbench (D*WB)*' 9th *WDK Workshop*, Rigi.

Andreasen M.M., Duffy A.H.B., MacCallum K.J., Bowen J. (1997) T. Storm, The Design Coordination Framework: key elements for effective product development, *The Design Productivity Debate*, A.H.B. Duffy, London, Springer-Verlag, pp. 151-172.

Borg J.C., (1999) *Design Synthesis for Multi-X,* PhD Thesis, Univ. of Strathclyde, Glasgow, UK June '99.

Borg J.C., Yan X.T., Juster N.P. (1998) A KICAD Tool for Pro-Active Exploration Support To 'Design Synthesis for Multi-X', *Proc. Of the 3rd International Knowledge Intensive CAD workshop* organized by IFIP 5.2, Tokyo Univ., Japan, 1-4 Dec. 1998, pp. 274-295.

CAMWorks™ (1997) from Teksoft™ CAD/CAM systems

Cunningham J.J., Dixon J.R. (1988) Designing with Features: The Origin of Features, *Proc. Of the ASME Computers in Engineering Conference*, San Francisco, California, August, pp.237-243.

De Kraker K.J., Dohmen M., Bronsvoort W. (1995) "Multiple-way feature conversion to support concurrent engineering", Proceedings of *ACM Solid Modeling '95*, Salt Lake City, USA, pp.105-114.

De Martino T., Falcidieno B., Giannini F., Hassinger S., Ovcharova J. (1994) Feature-based Modeling by Integrating Design and Recognition Approaches, *Computer-Aided Design*, Vol. 26, No 8, august, pp. 646-653.

Falcidieno B., Giannini F. (1989) Automatic recognition and Representation of Shape-Based Features in a Geometric Modeling System, *Computer Vision, Graphics and Image Processing*, 48, pp. 93-123.

FeatureWorks™ (1998) SolidWorks Corporation.

Henderson M.R. (1984) *Extraction of Form Feature Information from Three Dimensional CAD Data,* PhD Thesis, Purdue University, May.

Hubka V., Eder E. (1988) *Theory of Technical Systems: A total concept theory for engineering design*, Springer-Verlag Berlin/Heidelberg.

Krause F-L, Luddeman J, and Rieger E (1996) Feature based modelling of Conceptual Requirements for Styling, Proceedings IFIP WG 5.3, *Life Cycle Modelling for Innovative Products and Processes*, Berlin Nov/Dec 1995, F-L Krause and H. Jansen (Edts.) London Chapman & Hall.

March (1994) *A Primer on Decision Making – How Decisions Happen*, The Free Press New York.

Otto H. E., Kimura F. (1998) FT based systematic support of geometry related knowledge integration within product life cycle design. Pre-prints of the 3rd IFIP WG 5.2 Workshop on Knowledge Intensive CAD, Tokyo, Japan, Dec., pp. 258-273.

PART™ (1998) Tecnomatix Technologies Ltd.. from Van Houten, F. J. A. M. *PART: A computer Aided Process Planning System*, PhD Thesis, Univ. of Twente, 1991.

Pinnila J.M., Finger S., Prinz F.B. (1989) Shape Description and Recognition using an Augmented Topology Graph Grammar, *NSF Engineering Design Research Conference*, University of Massachusetts, Amherst MA, pp. 285-300.

Regli W.C., Gupta S.K., Nau D.S. (1995) Extracting alternative machining features: An algorithmic approach, *Research in Engineering Design,* 7(3), pp. 173-192.

Shah J.J. (1991) Assessment of Feature Technology, *Computer-Aided Design,* Vol.23 No. 5, June, pp. 331-343.

Shah J.J., Mantyla M. (1995) *Parametric and Feature-based CAD/CAM,* John Wiley & Sons, Inc.

Sreevalson P.C. (1990) *An Investigation into the unification of form feature definition methods,* Master Thesis, Arizona State University.

Tjalve E. (1979) *A Short Course in Industrial Design*, Newnes-Butterworths, London.

Tomiyama T., K.P Hew (1998), An Overview of the Modelling of Synthesis Project, *International Symposium on Modelling of Synthesis*, The Univ. Of Tokyo, Tokyo, Japan, The Modelling of Synthesis Project, Japan Society for the Promotion of Science (JSPS) & Research for the Future Program, pp. 3-12.

Vajna S., Wegner B. (1997), Support The Product Life Cycle By a Computer Aided, Integrated Feature Based Modelling Scheme. *11th International Conference on Engineering Design*, 19-21 August Tampere, Finland. Vol. 2 pp 301-306, Riitahuhta, A (Ed.) Tampere University of Technology Publisher.

Vandenbrande J.H., Requicha A.A.G. (1990), Spatial Reasoning for Automatic Recognition of Interacting Form Features, *Computers in Engineering*, Eds. G.I. Kinzel, S.M. Rohde, D.W. Bennett, C. Born, A.A. Busnaina, G. Gabriele, V. Ghazarian, J.J. Shah, T. Shoup, Book n.G0515A pp. 251-256.

Feature-Based Virtual Engineering

G. BRUNETTI
Centro de Excêlencia em Tecnologias Avançadas do Rio Grande do Sul - CETA-RS
Av. Assis Brasil 8450, 91140-000 Porto Alegre, RS, Brazil

Fraunhofer-Institut für Graphicshe Datenverarbeitung – IGD
Fraunhofer Straße 6, 64283 Darmstadt, Germany
gino.brunetti@igd.fhg.de
http://www.igd.fhg.de/igd-a2/

Abstract: In this paper the concept of Feature-based Virtual Engineering is presented as an approach to integrated product development. The focus of the paper is on presenting the current state of the development of a feature-based parametric representation for product semantics including a concept for creating and maintaining application specific views on the product information.

Key words: Virtual Engineering, Feature-based Product Representation, Integrated Product Development

1. INTRODUCTION

Nowadays, the use of virtual reality technology in engineering is mostly restricted to the design review and decision support. Typical VR applications are designed to provide an appealing visualisation of product geometry. Some of these applications support the shared viewing of virtual models in distributed environments. Others allow the animation of dynamic data obtained from, for instance, simulations and kinematics, or they support the simulation of processes like assembling. Emerging technologies like augmented reality for assembly and maintenance are only in a very early experimental phase. What these VR applications have in common is that the interaction with the model very often is limited to navigation and the release of predefined events. They usually do not allow the generation or modification of product information.

In order to achieve a further-reaching integration of VR in the product development process it is necessary to develop today's VR environments that are engineering data dead-ends into human centred interfaces of real Virtual Engineering (VE) environments. Such VE environments have to provide access to all available product information at the time required. The presentation of product information has to be configurable depending on the task that has to be performed. In addition to the traditional presentation of digital products VE environments have to allow to efficiently enter and modify product information. It must be possible to examine not only the physical aspects of products, but also to experience its function by transforming functional product information into corresponding interactive behaviour. Intelligent interaction methods have to be flexibly configurable depending on the user characteristics, the current task, the relevant information, and the I/O-devices available. Furthermore, VE systems have to provide intelligent feedback and guidance in order to support users as far as possible in achieving "correct" results. Finally, both the interaction control as well as the presentation of the product information must be automatically configured based on an interpretation of the product semantics instead of the manual scene configuration as it is usually done today. In this way, VE increases the intuitiveness of interaction with the digital product and raises the acceptance of the use of VR technology in engineering.

Feature-based VE environments as introduced in this paper are intended to overcome the disadvantages of today's approaches to an integrated product development that are based on data driven methodologies: the use of direct converters to exchange product data between highly specialised legacy systems; information lost or faulty information because interpretation of data is not stored within the product data structure, but interpreted by the various systems differently; insufficient support of modification back-propagation; and, integration only on file level. The introduced approach to VE in turn is based on an integrated product data structure that is feature-based and potentially able to store all relevant product information as well as to capture a high level of engineering semantics. In this approach product information is structured vertically according to different levels of semantic abstractions and horizontally according to the processes of the product development they belong to. Relationships between information units describing the different types of geometric as well as non-geometric properties of a product are managed via constraints and rules defining the knowledge-base about a product and all related processes.

For users the concept to access and manage product information is based on dynamically configurable views. Such views are defined by system configurations, by process information specifying the relevant product information, and by the filters and converters interfacing a VE environment with other systems producing and consuming the different types of product information. Furthermore, views are a means of abstraction allowing users to easily map their engineering know-how into a corresponding representation of product semantics in terms of meaningful features. Finally, it is the author's belief that feature-based virtual engineering as described above significantly improves the man machine interaction by providing highly sophisticated means of interaction with a digital product and all relevant meta data.

Figure 1. Elements of Virtual Engineering

The paper is organised as follows. First the basic idea of virtual engineering is described. Afterwards, the requirements on a feature-based architecture for a Virtual Engineering environment supporting different views on a product is presented. This follows an introduction into the used feature definition and a description of the feature-based product data model including views, which is also the basic content of this paper. Guided by the systematic approach to engineering design illustrated in *Figure 7* the discussion starts with the detail design phase followed by a bottom down approach from conceptual design to embodiment and assembly design.

2. VIRTUAL ENGINEERING

The implementation of the Virtual Engineering (VE) concept needs the integration of methods and tools, which are able of working in heterogeneous distributed system environments (de Amicis et al. 1999). In addition different visualisation techniques are used to support the different phases of the product development process in a goal-oriented way. The possible visualisation hardware spans completely different equipment such as the monitor in the office, a virtual table for working and management meetings, a large screen projection for the presentation of models and products in real-scale, or even a five sided CAVE.

The actual research activities in the field of information and communication technology lead to short intervals of technological innovations. A special interest can be found in the evolution of applications supporting simulation and visualisation of complex interactions in early states of the product design process so that the various aspects of a product can be analysed and optimised. In this context VE, for instance, allows the design review of products on basis of CAD data. Furthermore, process planning, production system planning and complex production facilities can be simulated, visualised and evaluated by applying VE technologies.

Besides these technology aspects VE concentrates on the relations between information technology and working environments. Therefore, a special focus is on the harmonisation of the technologies with the corresponding 3Ps: People, Processes, and Products as shown in *Figure 1.*

Nowadays, the product development is divided into several tasks related to the development competence of each partner. Process chains become more and more integrated and interdisciplinary. Therefore, VE applications must address the need of dealing with engineering teams where the members belong to different organisational units at distributed locations, where integrative processes supporting concurrent engineering even for interdisciplinary tasks are performed, and where various cultures regarding language, mentality, and working approach can be found.

The focus of this paper is the representation of product semantics that supports VE applications allowing users to interact and to experience the digital product according to the product information available along the product life-cycle. As outlined in the following sections the presented approach manages such product semantics within a parametric feature model.

3. FEATURE-BASED VIRTUAL ENGINEERING

It is generally agreed that computer supported engineering environments have to manage a much broader range of information than geometry. Besides other aspects it becomes more and more crucial to capture, represent and process the function and the functional behaviour of a product. This should already be done in the early phases of the product development process. Furthermore, designers should be assisted in developing syntactical correct and semantically consistent product models by providing intelligent and intuitive computer support that minimises the possible sources of engineering errors and maximises the capability to map designers' creativity into corresponding mathematical models of the product. The presented approach to these objectives is a feature-based representation scheme for product semantics including product function. Features are then the information carriers that allow modelling the relationships between requirements, functional descriptions and physical solutions of a product. They become the vehicle to bring this information to the downstream applications for embodiment, analysis, detail part design, and assembly such that it is possible, for instance, to keep track of the consistency between the concept, the design, and the manufacturing of a product. Approaches existing towards a feature-based integration of different phases of the product development to support concurrent engineering, which include previous work of the author, are based on extracting and mapping different views of a parametric representation of a product that are expressed in terms of geometric form features where products semantics are expressed in terms of explicit geometric or functional constraints (de Martino et al. 1998, Brunetti et al. 1996, de Kraker et al. 1995). Product semantics as it is handled within the conceptual design is not supported. Therefore, it is the aim of this paper to come up with a concept to extend these feature-based integration approaches in a way that it is possible to represent the conceptual design information, to support the evolution of product semantics along the product development process, and finally to improve concurrent engineering and top-down design. This is done by supporting an early feature-based prototyping of the different views to the overall product model making use of captured feature information to the largest extend possible. Major prerequisite of such an approach and therefore focus of this paper is a suitable representation scheme to store, manage, and retrieve product semantics and views on them.

3.1 Feature definition

The feature definition adopted in the presented work is that of the FEMEX (Feature Modelling Experts) working group established in 1996 by a number of international researchers (including the author), CAD system developers and larger CAD user companies (see, for instance, Vajna and Podehl 1998). According to the results of FEMEX a feature is defined as follows:

A feature is an information unit (element) representing a region of interest within a product. It is described by an aggregation of properties of a product. The description contains the relevant properties including their values and their relations (structure and constraints). Furthermore, it is defined in the scope of a specific view onto the product description with respect to the classes of properties and to the phases of the product life-cycle. Finally, a feature is described by properties out of several different classes of properties, thus relating these (classes of) properties to one another.

There are five major aspects in the definition above:

1. It is necessary to find a structure of properties suitable to express relevant product information in the different phases of the product life-cycle in terms of such product properties.
2. A feature is not limited to physical elements and exists only in the world of information models.
3. "Properties" are fundamental to the definition and at the same time the basic implementation mechanism.
4. The concept of views is inherent to the definition.
5. "Classes of properties" and "product life-cycle phases" are distinguished in the definition above. Some properties are meaningful in more than one phase, for instance, geometry – in different phenotypes – is considered in nearly every phase of the product life-cycle. The product information expressed in terms of properties aggregated by different features in different application contexts (views) is therefore the key mechanism towards an integrated product development.

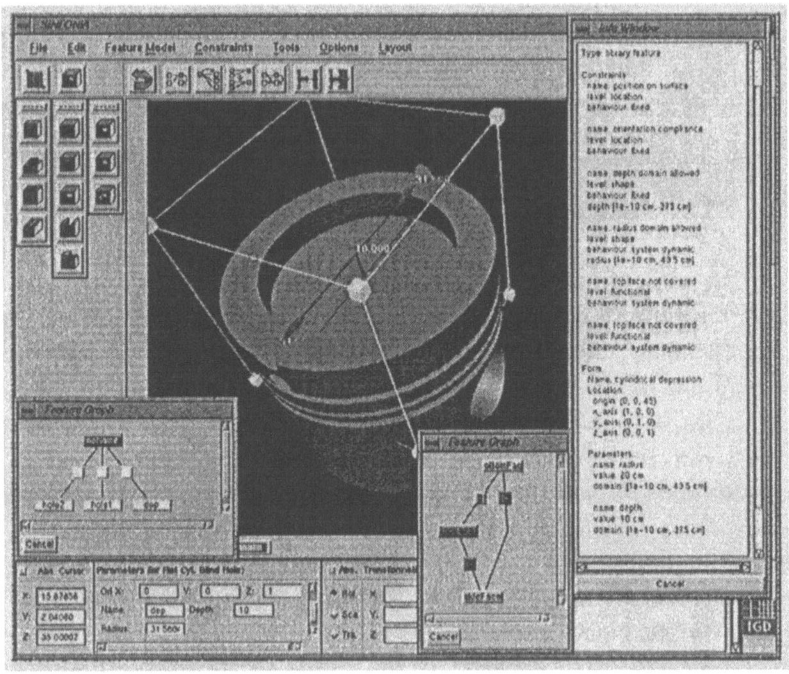

Figure 2. The feature-based parametric modelling system SINFONIA.

3.2 Feature-based parametric product Representation

The conceptual starting point and basis for the implementation of the presented approach towards an feature-based virtual engineering environment is the prototype system SINFONIA (see *Figure 2*). The latest results obtained with this system in the area of feature-based part and assembly modelling have been published in (Brunetti and Stork 1998). SINFONIA's underlying approach combines feature-based and parametric modelling methodologies and, following the FEMEX feature definition, structures feature data into four levels: an assembly model, a part model, a feature model, and a generic model. Starting with the assemblies every representation level is mapped on a model of the subsequent level, i.e. elements of a higher level are modelled by elements of the lower levels.

Application specific data describing a product assembly and its parts are specified in the first three models of the representation that consists of an assembly model, a part model, and a feature model. A feature model is built on top of the generic representation, which stores application independent product properties within a generic model. The data maintained in this generic model so far specifies the form and parametrics of the parts as well as the geometric and topological constraints defined on them. The final shape of the parts and their relative position within the assembly is derived from this generic model and mapped on a solid modeller.

3.2.1 Parametrics of the generic model

The generic model is an extended semantic net. Its nodes are topological form feature entities, geometric feature entities, or parameters. Arcs are constraints. It is a parametric model extending the declarative approach to feature-based modelling. Each form entity is always associated to its correspondent geometric description. Geometric entities are linked to each other by several types of spatial relations expressed as geometric constraints. The third type of entities are the geometric and functional parameters that can be constrained.

The spatial dependencies between features and their geometric entities are described by geometric constraints. Geometric constraints are, for instance, coincidence, complementarity, parallelism, intersection, perpendicularity, concentricity, and coaxiality. Parameters are controlled by domain and value constraints. Variations can be performed linearly, in equidistant steps, or by selecting values out of a concrete set. Interaction constraints specify the way form entities of features behave in relation to their intersection, attachment, and covering. Engineering constraints control the value and behaviour of parameters. They are expressed by analytic equations and build a system of non-linear equations. Using such constraints, parameter values are propagated, which implies the use of suitable solving techniques. Other engineering constraints are Boolean expressions which are used for checking logical conditions between parameter values.

3.2.2 Feature model

Application specific features and feature relationships build the feature model. Features contain their geometric, parametric, and functional description. They are related to generic objects representing the geometric and non-geometric properties of the feature. Feature parameters result from relationships established between different features of the model. Feature relations specify relative positions and parametric dependencies between features. They can be any non-empty aggregate of the supported constraints (see *Figure 3* and *Figure 4).*

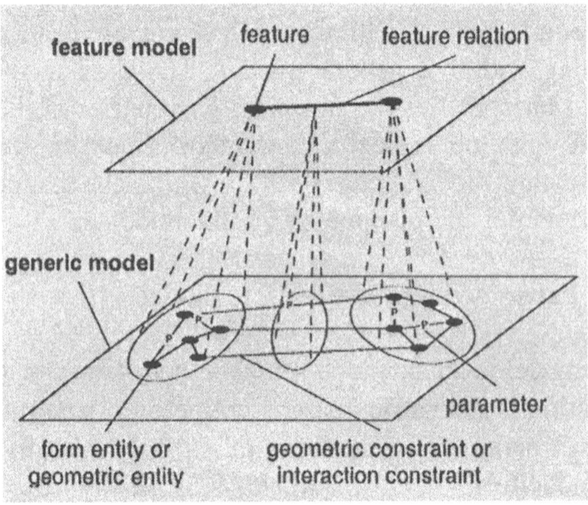

Figure 3. Mapping between a feature and its generic model.

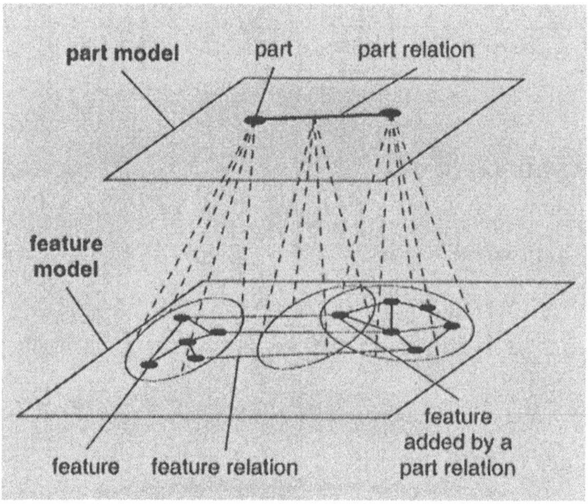

Figure 4. Mapping between a part and its feature model.

3.2.3 Part model

Parts, part parameters, and their relation-ships are kept in the part model that also builds a graph structure. A part node contains the functional description of a part from which the feature-based representation is derived. Part parameters are the parameters with a direct link to a corresponding feature parameter, but also part specific parameters as the material or the designer of a part. A part relation allows to specify a geometric or non-geometric detail of the relationship between two parts in an assembly. A geometric relationship between two parts is defined by instantiating a set of suitable feature relations between features of the parts (see *Figure 4*). If necessary, the features needed to establish the part relation are inserted in the feature model of the parts. Part relations then control the relative position of the parts and the dimensioning of the features involved in the part relations. Non-geometric part relations include, for instance, information about material compatibility's, needed tools, and installation space.

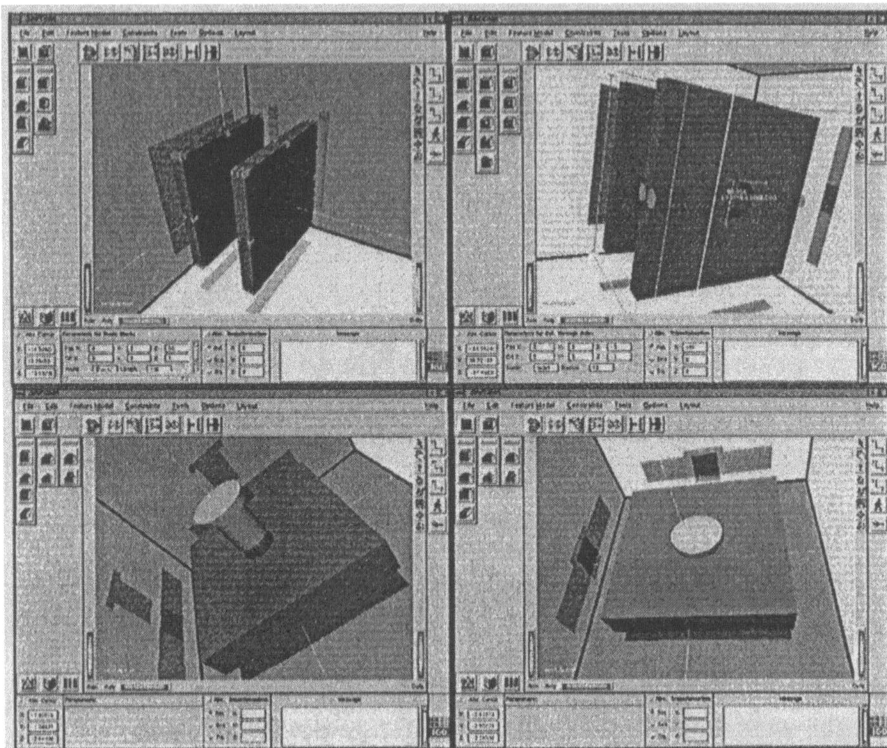

Figure 5. Attaching two plates by a bolt assembly within the SINFONIA system.

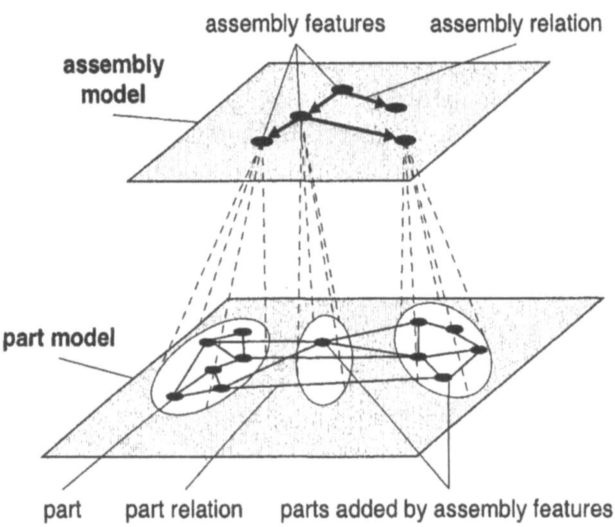

Figure 6. Mapping between an assembly and a part model.

3.2.4 Assembly model

The assembly model is organised as a tree with nodes representing assembly features and edges representing the assembly order. In addition to traditional assembly representations, assembly features contain not just the assembled parts or sub-assemblies of a product, but also the semantic of the relation between the parts. This semantic is part of the assembly feature description and includes the requirements on the parts to be assembled as well as all structuring information of the final assembly. Assembly features add part objects to the corresponding part model. These objects can be parts as, for instance, the screw of a screwing assembly, but are mainly a set of part relations. The relation between an assembly model and the subsequent part model is illustrated in *Figure 6*. See also *Figure 5*, which shows the current functionality of our prototype system SINFONIA. In the example two mechanical parts – plates – are connected by a mating part relation and by a bolt assembly feature that inserts two through holes to the model and creates a bolt to assemble the parts.

3.2.5 Feature-based conceptual design

Recently, the above described representation scheme of SINFONIA has been extended to the requirements of a more complete product data scheme extending the approach to the early phases of the design process, i.e. conceptual and embodiment design (Brunetti and Borut 2000). Following, the basic information units managed are shortly described and it is outlined how they have been inserted into the over all feature representation scheme.

All design starts with a requirements list that results from the very first task clarification phase (see *Figure 7*). It describes the general constraints a product has to or should fulfil. An important prerequisite to enable automatic validation of proposed solutions against given requirements is a structured and formalised description of these requirements that has to handle product function, effects, principal solutions, working principles, and finally also embodiments.

3.2.6 Product function

Technical systems perform the conversion of energy, material and/or signals. This conversion can be described as flow through the system, where one is the main flow and others, if any, are supporting flows. For the design of a technical system, a clear attitude between the input and output must be defined in form of the function of the system.

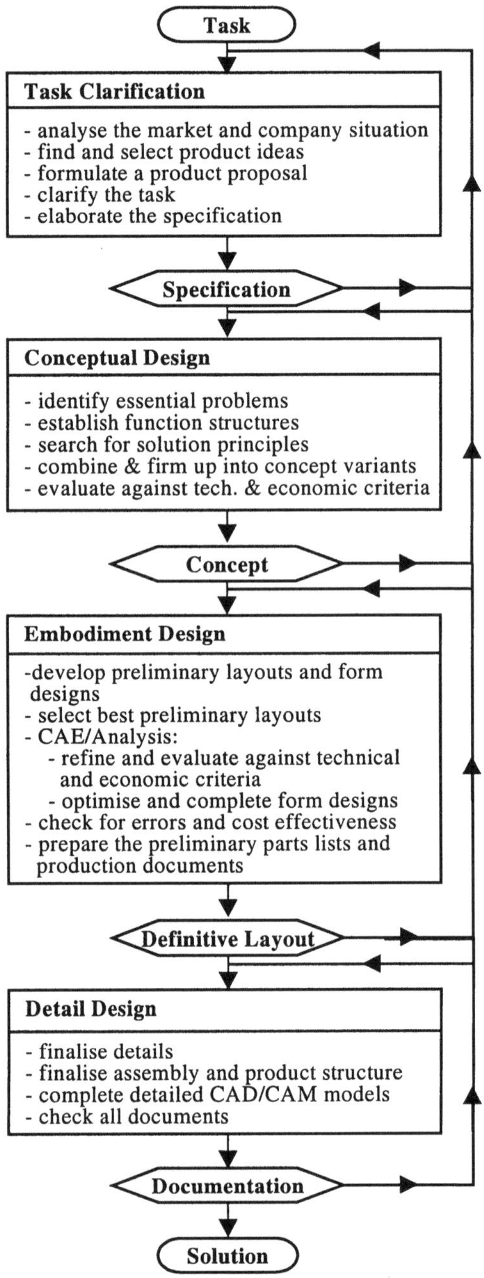

Figure 7. The design process according to the systematic approach to engineering design by Pahl and Beitz 1996.

This function is an abstract formulation of the task. The overall function can often be divided into sub-functions, and function structures can be established. With further sub-division of the sub-functions, basic functions - also called generally valid - can then be recognised. These functions, according to the definition by Rodenacker and Krumhauer (see, for instance, Pahl and Beitz 1996) are change, vary, connect, channel and store. They represent a conversion of type, magnitude, number, place and time, respectively. Functions are usually fulfilled by physical, chemical or biological processes, whereas mechanical engineering solutions are based mainly on physical processes. Selected physical effects and the determined material and geometric characteristics result in a working principle that fulfils each function. If a function cannot be fulfilled with a simple effect, a structure of effects has to be used instead. In the following phase of the design process - embodiment - qualitative and quantitative parameters such as surfaces, dimensions and material properties are specified according to the physical laws defining the effects.

3.2.7 Working principles

Figure 8 shows the mapping of a function structure into a working principle. To fulfil the main function of a product, an appropriate principle solution has to be selected. As described later, a principle solution is defined by an effect, an effect carrier, properties and physical laws. To perform the main function, auxiliary functions may be needed to supply supporting flows, to eliminate side effects or to meet given requirements (for example, in order to convert water into steam, an additional flow of energy is necessary). Auxiliary functions could be solved by the same or by additional principle solutions. Therefore, the functional structure has to relate sub-functions with principle solutions.

Physical effects are realised by the working geometry and by working motions. Working geometry is defined by arrangements of working surfaces. These can be varied with respect to and determined by type, shape, position, size and number. Type, nature, direction, magnitude and number determine working motions. Finally, basic material properties are defined by their state (solid, liquid, gaseous, space), their behaviour (rigid, elastic, plastic, viscous), and their form (solid body, grains, powder, dust).

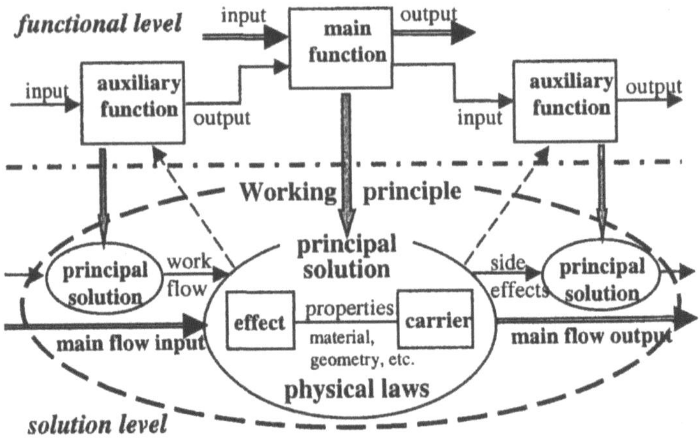

Figure 8. Function fulfilled by a working principle.

3.2.8 Function model

A feature-based representation scheme for capturing the product function must provide a means of explicitly modelling the sub-function structure as illustrated in *Figure 9*. Within such a function model, functions are represented by function features, which not only represent the static function information mentioned above, but also carry the knowledge about its intent and its concretisation in terms of principle solutions.

A principle solution consists of a chosen effect and the appropriate effect carrier. Usually, the same effect can generate several principle solutions, depending on material and geometrical properties; for example, the thermal dilatation effect can be combined with a solid body or a fluid as the carrier. The relation between function and principle solution is normally not one to one. Sometimes, one principle solution can solve several functions at once, however, a lot of functions cannot be solved by only one principle solution. In this case, a principle solutions structure is needed, which is the working principle.

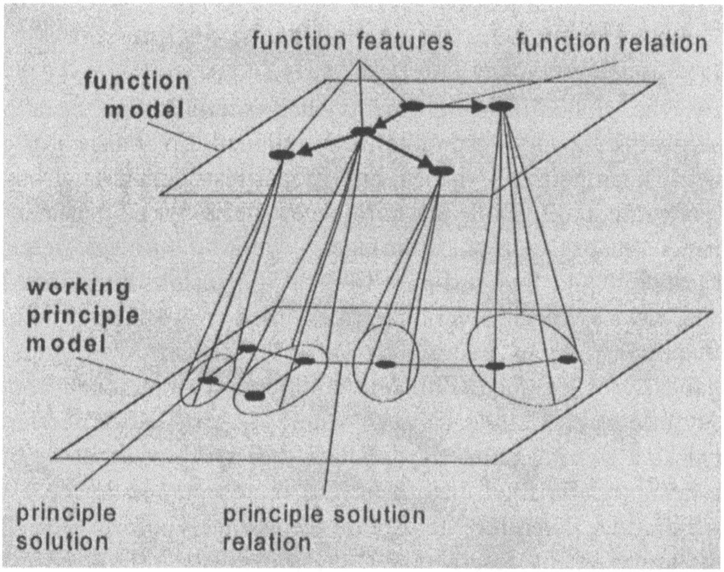

Figure 9. Relation between a function model and a working principle model.

3.2.9 Working principle model

As the number of known physical effects is deterministic, some authors have collected related principal solutions for use in the design process (Koller and Kastrup 1988). These catalogues of principle solutions are structured appropriately for a computer implementation (see, for instance, RPK 1999).

An appropriate structure of principle solutions, which fulfils the main function of a product, is a working principle. Together with given requirements, a working principle determines the embodiment of a product. Due to links/connections to auxiliary functions and a hierarchical structure of data, a working principle describes the principle solutions structure and implicitly also the function structure.

In the introduced representation scheme the principle solutions and their inter-relationships in terms of input/output parameters and their properties that make up a working principle are kept within a working principle model, where each principle solution is connected to the one ore more functions it realises. In this way it is possible to establish a parametric relationship between this working principle model and the function model as illustrated in *Figure 9.*

3.3 Feature-based embodiment design

Finally, the embodiment represents the materialisation of a concept where the overall layout of a product is determined. Embodiment is defined by geometrical properties - shapes, and properties of material from which part or parts are built. Both are influenced by a working principle and requirements. Shapes consist of surfaces - working surfaces, required by working principle and free surfaces. General dimensions are defined either by physical laws governing various effects used in working principles, or from requirements, or by material properties according to a strength and stress analysis, for instance. Principle solutions, as mentioned above, are defined by effects and effect carriers, where each effect carrier has to be considered as a starting point for detailing the layout of an assembly that, once the design is finished, has to have a representation in the assembly model of a product. Therefore, to capture the assembly information obtained during the embodiment design phase, the previously introduced assembly model is extended and related to the working principle solutions model as illustrated in *Figure 10.* The extensions are a means of representing the product structure and critical relations between assemblies instantiating the parametric relations between principle solutions, which are either basic functions or physical laws manipulating and controlling the input/output parameters of the functions and their principle solutions, respectively.

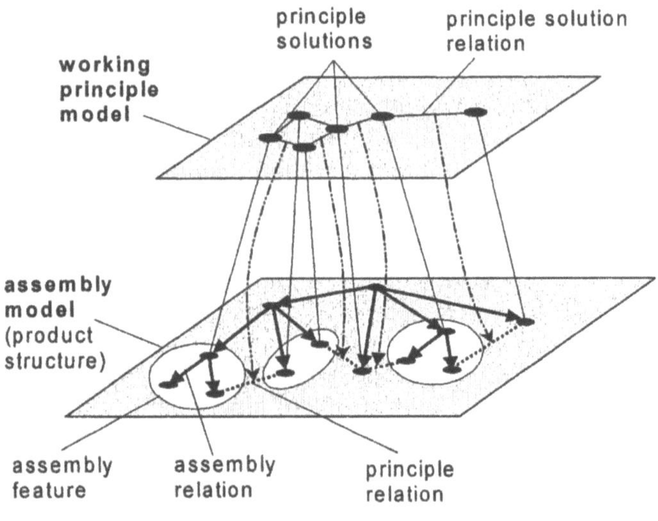

Figure 10. Mapping between a working principle model and an assembly model.

3.4 Views

From a certain perspective the different models defined in the previous section already define particular views onto product data. Nevertheless, views should be understood as a more general concept where any application used within the product development process defines its particular view to the product data. The goal is to come up with a mechanism that allows such applications to extract the required product information, to modify or create new product information, and to merge such information back to the all over product data in a consistent way.

An important prerequisite in order to enable the extraction of different application views to the product information is that this information is stored in small pieces, which are called atoms within this sub-section. Such atoms may be points and lines, but can be also free-form surfaces, a feature, a part, or a function. These atoms are related to one another forming a graph representing the basic knowledge about a product. Such a representation is called an associative or semantic net. The atoms are therefore nodes of a graph. A semantic net is a rather general means of representation. It is limited by the fact that only binary relations are allowed. E.g. the fact that five curves belong to a face of a solid cannot be directly expressed. This restriction is easily removed if general relations are allowed to be represented as atoms. Doing this systematically leads to the concept of conceptual graphs, which are semantic nets with two kinds of nodes. One kind represents the atoms, the other kind relations between atoms, such that the relations of the conceptual graph do not bear additional information. The current implementation of SINFONIA is based on the conceptual graph approach, which is in particular suitable for a complete object-oriented way of implementing both nodes and relations as classes.

However, a user normally does not enter atoms or relations between atoms directly, but she/he will use modelling operations at a higher level, which will generate many atoms and relations during one single operation (see, for instance, the description regarding the part relations). Even a simple modelling operation like „drilling a hole“ generates geometric atoms, topological atoms, functional atoms and possibly administrative atoms, which help to structure the information. Therefore, it is important that a user can look at and even modify that model in different views. A view is a way of structuring and displaying the information stored in the semantic net.

Now, by collapsing sub-graphs according to certain rules (graph-morphisms) different views on the product are provided. The resulting graph structures need not be explicitly stored in the system. E.g. it is not necessary to store an object representing a part with all its relations explicitly; the relations of a part should be consequences of the relations between the atoms that constitute the part. Thus the relations could at least in principle be calculated on the fly. It is left to the implementation, whether redundant information is stored or a calculation is performed. Features are identified with the morphisms mapping the atoms in the graphs, which describe the views.

Furthermore, views on the product data must be unique and have to be related to a specific phase (scope within the product life-cycle), version (modifications along the time) and configuration (alternative solutions at a given time) of the product data. This adds a requirement to the VE environment. This requirement is to support product data management functionality for all types and levels of product information. This finally also includes the views themselves.

More formally a view is a function mapping a subset of the product information (a specific version and configuration of atoms and atom relations within a specific scope defined by a phase) into a concrete product data representation that is itself part of the product data model.

The tools/algorithms to generate these views, i.e. map between concrete views, have to fulfil some requirements:

a) Mappings between views can be bijective, as in the case of mappings between different CAD systems, but they have to be at least surjective.
b) Conversions between views like CAD/CAE or also CAD/VR are partially injective, i.e. not all information is used for the conversion.
c) Filters are partially surjective, as they filter subsets of the product data model.
d) It should be supported to combine (cascade) mapping functions (A to B to C, if A to C is not defined or available).

Associativity between a concrete representation and the product data model is given, if in the case of partially injective mappings modifications on the data must be propagated back to the product data model and all other related views.

It follows that views should be managed within a model of views which itself is part of the product data model. Three major operations should be defined on this model (see *Figure 11*):

1. The generation of a concrete view out of the product data model, i.e. to generate a derived data model from the original data model.
2. The migration between concrete views, i.e. to map between derived data models (if necessary, via the original data model, i.e. the product data model).
3. The propagation of modifications between views, i.e. the update of the original product data model as well as the active derived (concrete) data models.

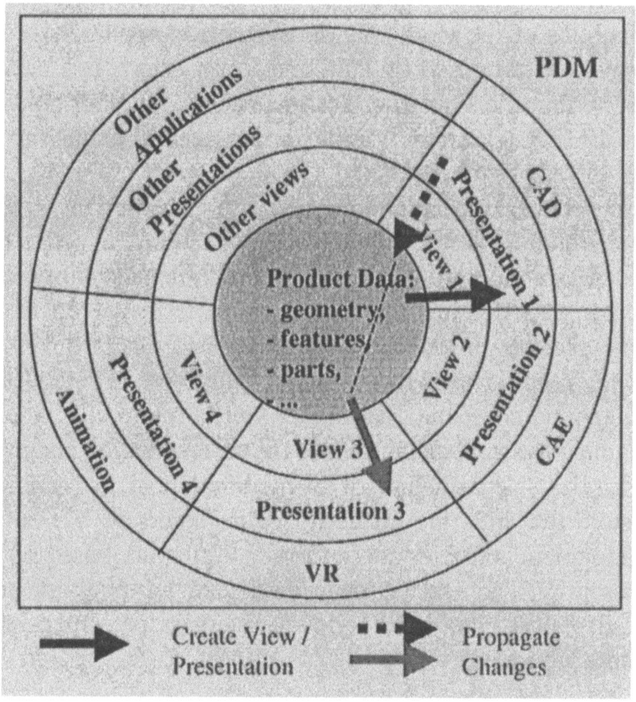

Figure 11. The concept of views.

The advantages of such an approach to keep a central product data model are that all applications of the VE environment work on the same database and that modifications performed by one application are visible to the other applications. Furthermore, by providing a mechanism to specify new mappings for views the system is open and extendable. Finally it takes into account the time dimension in case of product data revisions. The approach could be implemented, for instance, within a product data management system that provides functionality to manage versions and configurations, but on any atom level as presented above rather than on the CAD model level, as it is usually done today.

4. CONCLUSIONS AND FUTURE WORK

The work presented summarises the current results of the author's ongoing work towards feature-based Virtual Engineering. The current state of development regarding feature-based assembly modelling with SINFONIA has been presented. Furthermore, extensions of the product data model based on features to capture the relevant product semantics of the early design phases as well as a concept of views has been introduced.

Guiding principles of the work was the systematic approach to engineering design processes as proposed by Pahl and Beitz, which is a widely applied methodology within the European design schools. Another guideline of the work is the FEMEX feature definition as it is general enough to be applied for the scope of the presented topic.

The major concern for the time being is to efficiently represent the complex inter-relationships of products information in order to describe product semantics and views on them. Currently the possibility to model the presented parametric feature representation in terms of chain models on cellular complexes as introduced in (Palmer and Shapiro 1993) is being analysed, which could provide both an attractive formalisation and a mathematical sound computational model. Furthermore, this would allow to include dynamic physical behaviour into the representation and evaluation of product semantics. Nevertheless, it is the final goal to realise a human-oriented system that aids the creativity of designers in VE environments. Current development activities are in the area of feature-based styling of free form surfaces in VR, which has been motivated by the work presented in (Fontana et al. 1999). Finally, the view concept as presented in this paper will be further elaborated.

REFERENCES

de Amicis, R. Brunetti, G., Rix, J. (1999). *Virtual Engineering and Using Virtual Reality for Engineering*, Proceedings of the ADM'99, Palermo, Italy

Brunetti, G., Borut, G. (2000), *A Feature-based Approach towards an Integrated Product Model including Conceptual Design Information*, Computer-aided Design, Vol. 32 (14) pp. 877-887

Brunetti, G., De Martino, T., Elter, H., Falcidieno, B. (1996), *Modelling Shape and Semantics through an Intermediate Model*, Proceedings of the 29th Symposium on Automotive Technology & Automation (ISATA '96), Florence, Italy, pp. 71-81

Brunetti, G., Stork, A. (1998). *Product-centered Intituitive 3D Interaction for Feature-based Parametric Assembly Modelling*, Proceedings of the ProSTEP Science Days ′98, Product Data Technology Facing the Future

Fontana, A., Giannini, F., Meirana, M. (1999). *A Free Form Feature Taxonomy*, In Brunet, P, Scopigno, R.(eds.): Proceedings of the EUROGRAPHICS'99, Volume 18, Number 3

Koller, R. and Kastrup, N. (1998). *Prinziplösungen zur Konstruktion technischer Produkte*, 2. Auflage, Springer

de Kraker, K. J., Dohmen, M., and Bronswoort, W. Y. (1995). *Multiple way feature conversion to support concurrent engineering*, Proceedings of ACM Solid Modelling '95, Salt Lake City, Utah, pp. 105-114

de Martino, T., Falcidieno, B., Haßinger, S. (1998). *Design and engineering process integration through a multiple view intermediate modeller in a distributed object-oriented system environment*, Computer-Aided Design, Vol. 30, No. 6, pp. 437-452

Pahl, G. and Beitz, W. (1996). *Engineering Design, A Systematic Approach*, 2nd edition. London: Springer

Palmer, R. S., Shapiro, V. (1993). *Chain Models of Physical Behavior for Engineering Analysis and Design*, Research in Engineering Design 5, pp. 161-184

RPK [Institut für Rechneranwendung in Planung und Konstruktion, Karlsruhe(TH)]. Effektdatenbank [online]. (1999), http://rpkalf4.mach.uni-karlsruhe.de/~michelis/effects/

Vajna, S. and Podehl, G.(1998). *Durchgängige Produktmodellierung mit Features*, CAD-CAM Report Nr. 3, pp. 1-8.

An Approach to Implement Feature-Based Applications using Knowledge Aided Engineering Technology

F. MANDORLI[1], C. RIZZI[2], L. SUSCA[2] and U. CUGINI[2]

1 Department of Mechanical Engineering, University of Ancona, Via Brecce Bianche I-60131, Ancona, Italy

2 Department of Industrial Engineering, University of Parma, Viale delle Scienze 43100, Parma, Italy

Abstract: Concurrent Engineering methodology and process analysis and modelling showed the importance of tools that allow the designers to evaluate, from the beginning, different aspects of product definition. 3D CAD systems, simulation tools and virtual prototyping technologies can support the designer during her/his decision-making activity but, in order to provide an effective support, they must be tailored to the needs of the design process of each specific product.

During last years, different types of tools appeared on the market to support this tailoring activity and to promote the development of product-dependent applications. In this paper, we have focused the attention on software tools named KAE (Knowledge Aided Engineering) development shells. Applications implemented by using this type of tools can support experts during the decision making process to evaluate alternative solutions. They provide a kernel for the integration of different technologies, and represent a basic step for the development of the product design.

In this paper we present a brief overview of the architecture of a typical KAE application and we provide a survey of the most significant systems that we have developed, within different application domain, taking advantage of the KAE technology.

Key words: Knowledge Aided Engineering, Feature Technology

1. INTRODUCTION

The development of ideas and design conceptualisation involve creativity, expertise (Cross, 1998); at the same time, to reduce design cycle time and improve product values, the designer must be able to identify the best design solution as soon as possible, i.e., the *right design the first time.* A crucial issue in decision-making process is the possibility to design and evaluate alternative solutions in a short time.

The problem to compute and verify alternative design solutions is a traditional engineering problem that involves routine jobs and manual computation and often requires an enormous waste of time (Vernadat, 1996). In this scenario, simulation can play a significant role by providing simplified (either physical or mathematical) models the designer can use to validate his/her assumptions.

The motivation for the development of an automatic decision support system is to eliminate routing jobs and perform, as much as possible, automatic computation during the evaluation of alternative design solutions. This will speed up the decision process and will give to the designer time to develop new ideas and improve the product.

A key aspect in the development of such a kind of systems is the amount and the quality of controls incorporated into the system, i.e., in an ideal situation, the user should be able to interact with the system in order to generate alternative design solutions, and the system should give feedback about the qualitative and/or quantitative evaluation of each solution.

In order to provide such functionality, the system must incorporate all the significant knowledge about the problem under inspection. This objective can only be reached through the formalisation of related know-how and consequent acquisition and mapping of knowledge into appropriate computerised models and functions.

A possible methodology to deal with the problem of knowledge representation within design support systems is based on the concept of feature. The concept of feature provides the basis to formalise and represent the different types of knowledge involved in the designer's decision-making activity. In fact, features have been widely used to represent complex structures containing shape as well as functional and technological information of a specific application domain (Shah 1994, Shah 1995) and the feature technology has matured to the point that most of the today available CAD systems claim to be feature-based.

Although feature-based systems have improved the modelling capability of CAD systems, the problem to define general procedures able to guarantee the control of the coherence between function and shape (i.e., the feature semantic) is still an open issue (Bidarra 2000, Gao 2000, Mandorli 1997).

Better results on model control can be achieved when the general purpose approach is substituted by a domain-dependent approach, i.e., instead of using a general purpose *shape-based* system, we move to a *function-based* system, tailored on a specific product and incorporating the users' knowledge related to the different phases of the product design process. In this case, the problem to tailor a generic system to the needs of a specific application domain is strictly related to the definition of an appropriate set of features and to the implementation of suitable functionality to manage and control the feature model.

Several approaches based on different technologies and with different level of complexity, are today available to face this problem.

In this paper we present three different applications we have developed by using a feature-based approach supported by the Knowledge Aided Engineering (KAE) technology.

The intent is to summarise our experience and to underline advantages and drawbacks related to the adopted technology, with reference to the identification of class of design problems that can benefit from the proposed approach.

2. TECHNOLOGY TO DEVELOP TAILORED APPLICATIONS

Different IT tools can be used to implement customised applications. First consideration when selecting the most appropriate development tool is the needed balance between the amount of control that will be incorporated into the system and the development effort.

Remarks about the effort and skill required to develop a tailored application is particularly significant if we consider the lost of information that may occur when the domain expert has to interact with the IT expert that will develop the application. This drawback could be avoided if the domain expert can develop, test and finally use the application.

In the following we summarise the approaches to the development of tailored applications, based on different technologies, and we give an assessment about the control/development-effort balance (See Table 2).

- **CAD Feature Libraries**:
  most of the commercially available feature-based CAD systems allow the users to create libraries of feature templates. The feature template is generated by defining its shape and main geometric parameters. The user can interactively define new features and store them in a feature library

for a future use. The control of the correct use of the feature template during the definition of a new model is left to the user.

- **CAD Development Kit**:
most of the commercially available CAD systems provide a software development kit. The developer to implement control algorithms that will navigate the data structure in order to check the model consistency can use this kit. The developed algorithms become available as new commands in the system interface and the user can recall them when required.
- **Software Components:**
several software components are available on the market as software libraries to provide specialised services for geometric modelling, dimensional constraints management, visualisation, etc.. The developer can take advantage from the algorithms present into the libraries to develop its own application.
- **KAE Development shell**:
the user can develop his own application by using a development environment that provides functionality to handle a hierarchical product model structure including shape representations, dimensions, assembly rules, access to external data bases, etc.

Table 2. Assessment of the balance between development effort and model control

	Development Effort	Control
CAD Feature Library	The user interactively defines the features. Features with multiple shape representations must be modelled as separated features.	Delegated to the feature parameters. The correct use of the feature is left to the user.
CAD Development Kit	The developer takes benefit from CAD system data structures and algorithms to implement additional control commands.	The user from the CAD system interface fires the new developed control command.
Software Components	The developer must define appropriated data structures and fully implement the system by taking advantages from the services provided by the software libraries.	The developer has to implement the controls that will perform automatic check or will be available for the final user.
KAE Development Shell	The developer must define the product model in terms of parts, subparts, part properties, and parts relationships.	When a model property is changed, the developed application automatically re-computes the model on the basis of the controls inserted into the model in form of part properties.

The CAD feature library approach does not require a real implementation effort; however, the features defined in such a way can just represent shapes and related dimensioning parameters, but they fails in the possibility to manage validation rules and other complex functionality to control the correct behaviour of the feature model.

The CAD Development kit and Software Components approaches allow implementing better control procedures but they require the developer to have a significant skill in both programming languages and geometric modelling technology.

KAE development shell represents the compromise among the above-mentioned approaches. The developer can take benefit from the flexibility of the high level programming language provided by the tool to define a product model containing all the different types of required knowledge. This approach requires a limited experience in programming and geometric modelling technology. In next chapter we will describe with more details the characteristics of a KAE development shell and the methodology to implement an application.

3. KAE DEVELOPMENT SHELL

3.1 Methodology to develop KAE applications

KAE development shells are based on *generative technology.* Within the development environment, the domain expert can create and test generative product models. Once the model has been completely defined, a graphical user interface for the application has to be developed in order to let the end-user interact with the generative product model (Rosenfeld, 1995; Moulianitis, 1999; Ognjanovic, 1996).

The generative product model is a hierarchical structure made by parts and subparts. Part properties and relationships have to be defined by the domain expert in order to incorporate design specifications, design intent, engineering and manufacturability rules, and enterprise-wide best practices.

The generative product model takes, as input, the design specifications, applies relevant procedures, and generates automatically a product design. When the design specifications change, the KAE shell automatically updates the product model, and generates new design version by directly deriving all the outputs.

By means of the user interface, the end-user can change the design specifications and execute the generative model over and over again to rapidly evaluate alternative design solutions, taking advantage from the automation of several routine operations, such as the following ones:

- Generating geometric models from functional specifications;
- Calculating engineering and geometry properties;
- Choosing a configuration of product components;
- Enforcing compliance to standards and best practices;
- Verifying product manufacturability;
- Determining the theoretical performance or behaviour of the product;
- Exporting the product geometry and model information to external engineering analysis programs;
- Generating information such as bills of materials, cost reports, manufacturing instructions and CAD-format data.

3.2 Components of a KAE development shell

The main components of a KAE development shell include:

- The interpreter of the generative design language used to define product models;
- The dependencies manager to trace product model changes and drive the model re-generation;
- Geometry modelling tools to create rule-based definitions of surface and solid models;
- Graphical interfaces to interact and test the generative product models;
- Data integration tools for linking to other software products.

The core of a KAE shell is the interpreter of the generative design language and the dependencies manager that drives the model re-generation when model inputs change.

Regardless syntactical aspects, the generative design languages provided by the different commercially available KAE development shells are based on high-level Object Oriented languages with multiple inheritances integrated with functionality to represent part-subpart relationships.

Each part of the generative product model is defined as an object having a name and a set of property/expression couples representing respectively the name and the value of the property. The property value is an expression (i.e., any valid sentence of the language). New objects can inherit properties and property values from basic objects or from previously defined object. Sets of basic objects are supplied with the shell in order to provide basic functionality. These objects can be roughly grouped into following categories: objects to handle series of identical objects, objects implementing methods to access external databases, objects implementing calls to the integrated geometric modeller, and objects to handle input/output functionality.

By using a particular part-composition property, the objects can be organised into a tree structure in order to form the generative product model.

The language provides a special syntax that can be used in any expression to make reference to an object property. This feature, together with a dependencies manager, allows the developer to build parametric models: when a property is evaluated by computing its expression, if the expression contains references to other properties, the dependency manager automatically and recursively evaluates all the properties which are referred to.

While defining mechanical product models, shape is a significant part of the information that must be used to represent product components. In this case, the developer can benefit from the geometric objects provided by the KAE shell. Once a new object inherits from a basic geometry object, it inherits all the methods that are needed to render the object and the properties that are needed to represent the basic parameters of the object shape. Basic geometry objects can be combined to form more complicated shapes. Using expressions referring to other properties of the model can parameterise the dimensions of the geometric objects.

When the product model has to represent an assembly, a further need is the possibility to position a model component with respect to another one. This can be done by means of positioning property defined within the geometric objects. A suitable syntax allows to define positioning constraints such as: align, mate, angle, centred at, coincident, concentric, match vectors, parallel, perpendicular, tangent, etc.

4. EXAMPLES OF DEVELOPED KAE APPLICATIONS

In the following three different applications developed by using a feature-based approach supported by the Knowledge Aided Engineering (KAE) technology are presented.

4.1 Automatic generation of wire harness design and schematics

The application described in this section is an automatic system to support design and comparative evaluation of different wire harness architectures. The application simulates the traditional design process and still leaves a fundamental role to the designer who makes the choices.

Wire harness design is an evolutionary process, which transfers the logical model of required electric functionality from the abstract domain of functions and logical nets to the physical domain of connectors and cables.

The logical model is given in terms of circuit diagrams where all required functionality are represented together with all the logical connections among them.

While designing wire harness for automotive industries, the decision process of how to map the logical model into a physical model is basically constrained by the vehicle layout, materials, manufacturing techniques, and economics.

Depending on how functionality are grouped into connectors and uses, the mapping process can lead to solutions having different wires length, type and even number.

The developed application is based on a product model structure made by a set of features organised into four main subparts representing the four logical class of information to be managed: the entire set of circuit diagrams that are part of the project, represented in terms of electrical symbols and logical nets; the whole set of cabling, grouped by regions, represented in terms of tubes and components; the set of automatically generated wire harness schematics represented in terms of components and wires; the vehicle layout, represented in terms of different vehicle regions (engine, dashboard, door, etc.) and the set of tubes connecting different regions.

During a typical working session, the system takes as input a set of files, representing circuit diagrams, generated by an electrical 2D CAD system. An interactive session is then required where the user, supported by a graphical environment, must specify the connector and tubes location. Starting from such information the system generates a wire for each logical net that is part of one of the diagram. The components to be connected by the wire are evaluated by looking at the correspondence between logical symbols and physical components. A graphical representation of the wire harness is then provided (see *Figure 1*) together with information about wires length and type, evaluated by looking at the tubes path the wire must follow to connect the components.

By changing the connector location and/or the tubes the user can generate and evaluate alternative solutions.

The application has been developed in collaboration with CRF (Italian Research Centre with participation of Fiat S.p.A.) and IVECO (the Italian leading company for truck production). The application has been realised using the ICAD development shell by KTI Inc. (Di Lecce 1997).

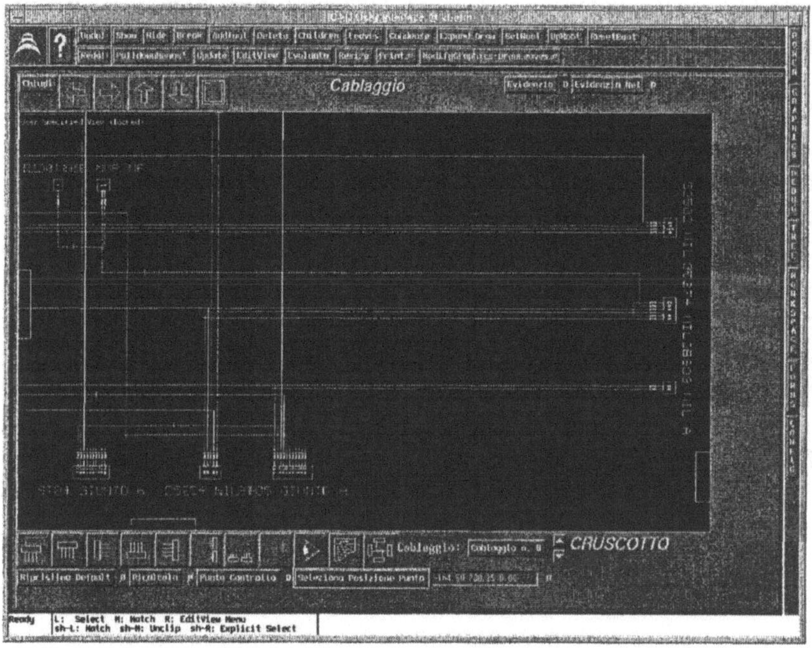

Figure 1. Example of Automatically Generated Wire Harness

4.2 Weights distribution analysis of IRL Cars

The application described in this section is an automatic system that allows the designer to automatically compute and evaluate mass properties of Indy Racing League (IRL) cars.

During the design process, car components are sized in order to optimise the weight/strength balance with respect to the overall constraints imposed by the IRL technical specifications. Car mass distribution is then derived from the dimensioning of the car sub-assemblies. However, mass distribution is a fundamental parameter, which must be taken into account during design because of its contribution to car performance. The designer tries to improve car performance starting from the idea of a new car configuration. In this stage, the designer is guided by the perspective of achieving some particular advantages that s/he can foresee relying on personal expertise. Possible alternative solutions are evaluated in order to define the strategy of the following steps. A detailed analysis of the solution selected allows the designer to evaluate his/her original idea, giving a feedback of results on the hypothesis

The developed application is based on a car model made by a set of features organised into a product model structure containing all car sub-systems (e.g., suspension and chassis). Different part-subpart relationships define the tree model and link an object (e.g., suspension) to its components (e.g., wishbones and wheel).

Starting from a set of predefined independent parameters (design variables), the system automatically generates the car model configuration, including geometric models used to evaluate all properties related to derived dimension, position, mass, etc.

During a typical working session, the user interacts with the graphic user interface (see *Figure 2*) to edit independent design parameters and generate alternative car configurations. The user evaluates different car configurations by looking at the system outputs, which consist in numeric data (mass, centre of mass of both the car and its sub-systems) and graphic elements (car and sub-systems 3D representation).

The application has been developed in collaboration with Dallara (Italian leading company in the production of racing cars) using the Selling Point development shell by Oracle Corporation (Susca 1999).

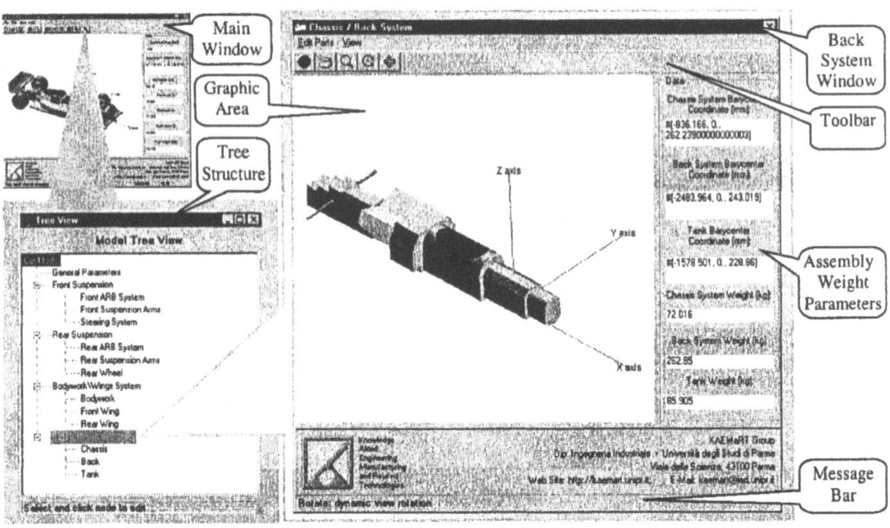

Figure 2. User Interface of the Weights Distribution Analysis System

4.3 Cupboard configuration

The application described in this section is an automatic system for cupboard configuration that drives the user to the definition of valid configurations and automatically generates bills of materials, NC programs and offers for sale.

Furniture configuration is traditionally performed by selecting and composing standard components (wall, shelf, back, drawer, door, etc.) from a catalogue. Due to assembly and manufacturing constraints, only a subset of all the possible configurations is allowed. The control about the validity of a configuration is usually left to the experience of the expert, usually sales assistant. If a configuration error occurs during the selling, this is usually detected during the assembly of the furniture. The error produces the need to reconsider the configuration and then to make corrections on the bill of material, with a consequent changes on the final price. This situation will delay the delivery and the final assembly of the furniture and will cause the customer disappointment. The risk to make a mistake during the configuration process rises when the customer asks for the modifications of standard components in order to meet constraints imposed by the furniture location (room dimensions, steps on the floor, etc.).

The developed application is based on a cupboard model made by a set of features organised in two main subparts representing the internal and external components of the cupboard.

The external part of the cupboard has a predictable structure made by a sequence of aggregated components and then it can be automatically generated starting from a set of independent parameters, such as the number of cupboard aggregates and the dimensions of each aggregate (see *Figure 3*).

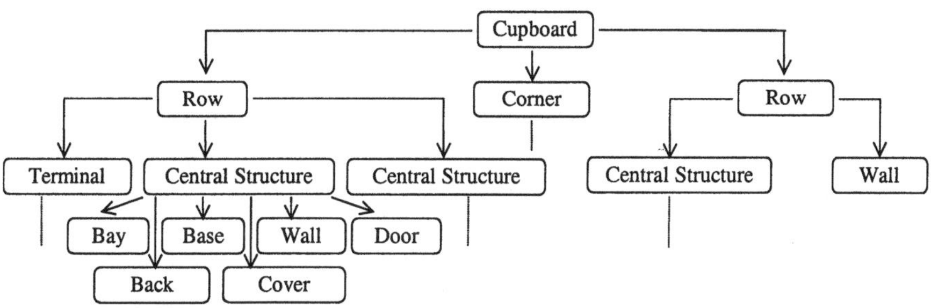

Figure 3. Model Structure of the Cupboard Exterior

Conversely, the cupboard interior is dressed by interactively selecting a variable number and type of components that can be assembled on the basis of specific assembly constraints. In order to manage this situation, the subpart structure representing the cupboard interior has been enriched with a logical feature (called bay) that simulates the internal available space. The bay feature behaves as follows: when a new component is candidate to be inserted in the space available in the interior model structure, the bay feature checks the compatibility between the selected component and the selected location. This can be done thanks to the knowledge the bay feature has about the type of components surrounding the available space and the assembly constraints. Once the selected component has been verified and accepted, it is inserted in the interior model structure together with two new bay features, resulting from the split of the original one (see *Figure 4*).

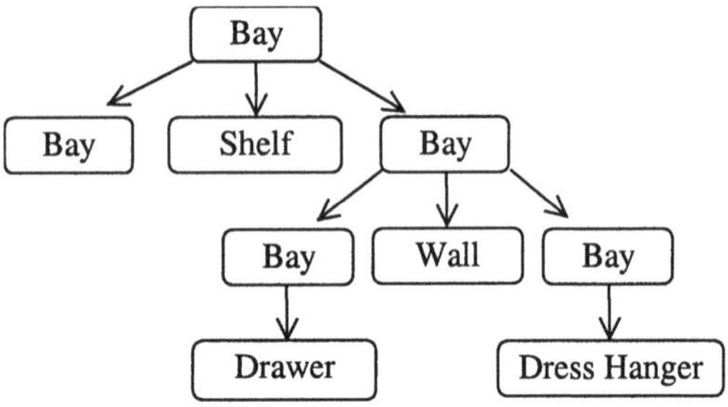

Figure 4. Model Structure of the Cupboard Interior

During a typical working session, the user first interacts with the graphic user interface to generate the cupboard external structure by providing independent parameters, such as number and dimensions of the cupboard aggregates (see *Figure 5* top). Then s/he starts to dress the cupboard interior by selecting internal components (see *Figure 5* bottom). Once the configuration process is completed, the user can ask the system to automatically generate bill of material, NC programs for the components that requires additional manufacturing, and sale offer.

The application has been developed in collaboration with Molteni (Italian leading company in the production of furniture) using the Concept Modeller development shell by Wisdom Systems (Mandorli 1995).

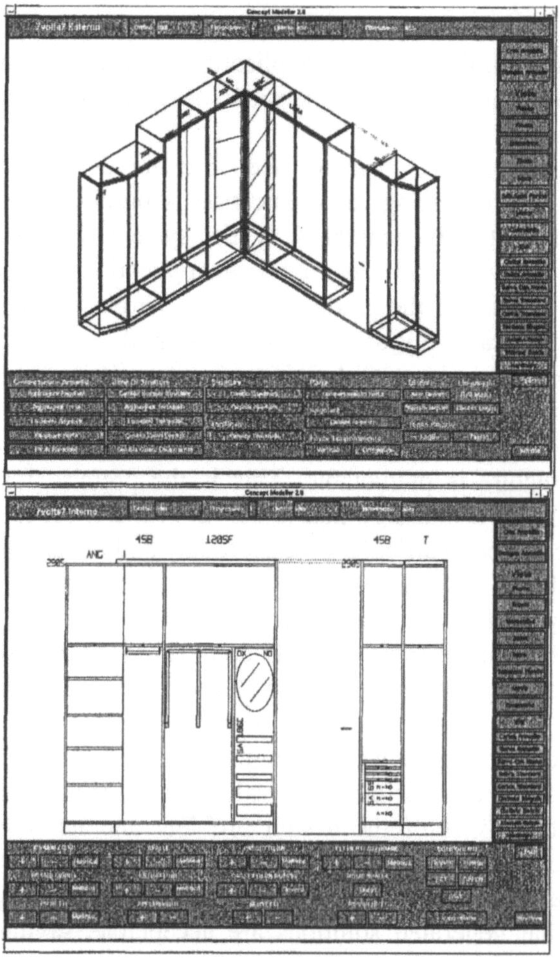

Figure 5. User Interface of the Cupboard Configuration System (for exterior and interior definition)

5. CONCLUSIONS

The paper describes three different applications developed with the support of KAE technology.

Several distinctions can be noticed among these applications, regarding geometric aspects, product structure, product definition and evaluation process. In particular, for the wires application the geometry representation required to support the evaluation process is mainly 2D (automatic generation, representation and positioning on a plane of a huge number of wires), while both for racing car and cupboard application it is 3D. The IRL car application requires the evaluation of mass and centre of mass properties of the car 3D model, the cupboard application requires reasoning about the adjacency relationships and the dimensions of the 3D objects that build-up the cupboard model.

Regarding the product structure and structure definition process, the product model for the racing car application is purely generative (the car structure is automatically generated by the system on the basis of user inputs), while the wire and the cupboard applications require an hybrid approach where a part of the product model is automatically generated and another part is interactively defined by the user: the truck cabling for the wire application and the interior definition for the cupboard application.

Nevertheless, a significant common point can be identified among all of the three developed applications: the design process is known and stable and the product structure is predictable. This is the premise for a successful application of KAE technology that is based on the exact identification of the design process for a particular product family and on its organisation into a prototype, which is characterised, by the same rules and procedures that are traditionally used by the designer during her/his activity.

The most critical issue of a KAE application is the analysis of the design/manufacturing process, that must lead to the identification of the independent design parameters (later used to generate alternative design solutions), the constraints each solution must satisfy, and the criteria that will support, or automatically perform, the evaluation of the generated solutions.

The correct analysis of the design problem and its mapping into an appropriate product model can lead to the development of a system able to perform all the repetitive tasks of the design process and to automatically verify/ensure that generated solutions are acceptable, saving a lot of the designer's time.

A significant side-effect of the development of a KAE application is that the implemented product model serves as a centralised repository for engineering information and expertise, that can be eventually re-used for education/training of junior engineers.

ACKNOWLEDGMENTS

This work is dedicated to our friend Lorenzo.

REFERENCES

Bidarra, R., Bronsvoort, W. F. (2000), Semantic Feature Modelling, Computer-Aided Design, Elsevier Science Ltd., N. 32, pp. 201 - 225.

Cross, N. (1998). Expertise in Engineering Design. Research in Engineering Design, Vol. 10, pp. 141 – 149.

Di Lecce, B., Tagliabo', F., Mandorli, F., Accornero, B., Strata, A. (1997), A Knowledge-Based tool for automatic generation of wire harness design and schematics, in Proceedings of 6th European Conference, Lightweight and Small Cars, the Answer to Future Needs, Vol. 2, pp. 987 - 998.

Gao, S., Chen, Z., and Peng, Q., (2000), Feature Validity Maintaining Based-on Local Feature Recognition, In CD-ROM Proceedings of the 2000 ASME International Computers and Information in Engineering Conference, ASME.

Mandorli, F., (1995), Sistemi per lo sviluppo di applicazioni KAE: un esempio nel settore degli armadi componibili, In IX Convegno Nazionale Associazione Nazionale Disegno di Macchine (ADM '95), pp. 171-180.

Mandorli, F., Cugini, U., Otto, H.E., and Kimura, F., (1997). Modeling with self Validation Features, in Proceedings of ACM/IEEE Symposium on Solid Modeling and Applications '97, pp. 88-96.

Moulianitis, V. C., Dentsoras, & A.J., Aspragathos, N.A. (1999). A Knowledge-Based System for the Conceptual Design of Grippers for Handling Fabrics. Artificial Intelligence for Engineering Design, Analysis and Manufacturing (AIEDAM), Vol. 13, No. 1, pp. 13-27.

Ognjnovic (1996). Decisions in Gear Train Transmission Design. Research in Engineering Design,Vol. 8,pp. 178 – 187.

Rosenfeld, L.W., (1995), Solid Modeling and Knowledge-Based Engineering, Handbook of Solid Modeling, Eds. Donald E. LaCourse, McGraw-Hill Inc.

Shah J.J. and Mäntylä, M. (1995), Parametric and Feature-Based CAD/CAM: Concepts, Techniques, Applications, John Wiley & Sons, New York, NY.

Shah, J.J., Mäntylä, M. and Nau, D. (1994), Introduction to Feature Based Manufacturing, Advances in feature based manufacturing, Elsevier Science B.V.

Susca, L., Mandorli, F., Rizzi, C., and Cugini, U. (2000), A Racing car design using knowledge aided engineering, Artificial Intelligence for Engineering Design, Analysis and Manufacturing (AIEDAM), Vol. 14, No. 3, pp. 235 – 249.

Vernadat F.B. (1996). Enterprise Modeling And Integration – Principles and Applications. Chapman and Hall.

Product Development with Multiple-View Feature Modelling

W. F. BRONSVOORT, A. NOORT, J. van den BERG and G.F.M. HOEK
Computer Graphics and CAD/CAM Group
Faculty of Information Technology and Systems
Delft University of Technology
The Netherlands
W.F.Bronsvoort/A.Noort@its.tudelft.nl

Abstract: Multiple-view feature modelling can support applications from various phases of product development, by providing an own view on a product for each of these applications. Each view contains a feature model of the product specific for the application. Current approaches to multiple-view feature modelling only deal with form features. Hence, only the later product developments phases, in which the geometry has to be fully specified, are supported. In addition, they only deal with single parts, not with assemblies. A new approach to multiple-view feature modelling is presented here that overcomes these shortcomings. It supports conceptual design, assembly design, part detail design and part manufacturing planning, by providing a view for each of them. The views and the operations that can be applied on them, and the ways the views are kept consistent, are described.

Key words: Feature modelling, Feature views, Consistency maintenance, Conceptual design, Assembly design, Part detail design, Part manufacturing planning

1. INTRODUCTION

Contemporary product development requires dealing with resources economically, and aims at higher quality of products. To realize this, concepts like concurrent engineering and feature modelling have been introduced. Concurrent engineering aims to design better products in less time, by using Design For X (Huang 1996) and simultaneous engineering

(Bullinger and Warschat 1995). Feature modelling aims at providing ways to specify more design intent in a product model.

In feature modelling, the model of the product to be developed is built from features, which can be described as entities with some attributes and properties that are useful in the reasoning process of one or more applications (Shah and Mäntylä 1995).

Multiple-view feature modelling is a concept that combines concurrent engineering and feature modelling. It supports applications from various phases of product development, by providing an own interpretation of, or view on, the product for each of these applications. Each view contains a feature model of the product specific for the application, which is built from instances of generic features that are stored in the feature library of the view (Bronsvoort et al. 1997, De Martino et al. 1998, Hoffmann and Joan-Arinyo 2000).

The main shortcomings of the current approaches to multiple-view feature modelling are that they only deal with form features, i.e. regions of the product geometry that have some functional meaning, and hence support only the later product development phases in which the geometry has to be fully specified, and that they only deal with single parts, not with assemblies. In this paper, a new approach to multiple-view feature modelling is described that overcomes these shortcomings, by also providing support for conceptual and assembly design.

Section 2 describes some existing approaches to multiple-view form feature modelling. Section 3 presents the new enhanced multiple-view feature modelling approach. Sections 4, 5 and 6 describe the way conceptual design, assembly design, part detail design and part manufacturing planning are supported by the new approach. Section 7 presents the way the feature models of different views are kept consistent, e.g. when the feature model of one of them is changed, and Section 8 gives some conclusions.

2. MULTIPLE-VIEW FORM FEATURE MODELLING

Several approaches to managing multiple form feature views exist. With feature conversion, or to be more precise feature model conversion, a new feature model for some application is derived from an existing feature model for another application, and changes made in the feature model for one application are propagated to the feature models for other applications.

In one-way feature conversion, typically, the feature models of analysis and planning views are derived from the feature model of a design view. If the analysis or the planning results in the need to change the model of the product, then these changes have to be made in the feature model of the design view, and the feature models of the other views have to be updated from that model.

Cunningham and Dixon (1988) propose a multiple-view system architecture with one-way feature conversion. In this proposal, the designer uses design by features to build the feature model of the design view, called the primary view, and feature conversion is used to derive the feature models of, for example, the finite-element view and the process planning view, called secondary views (see *Figure 1*).

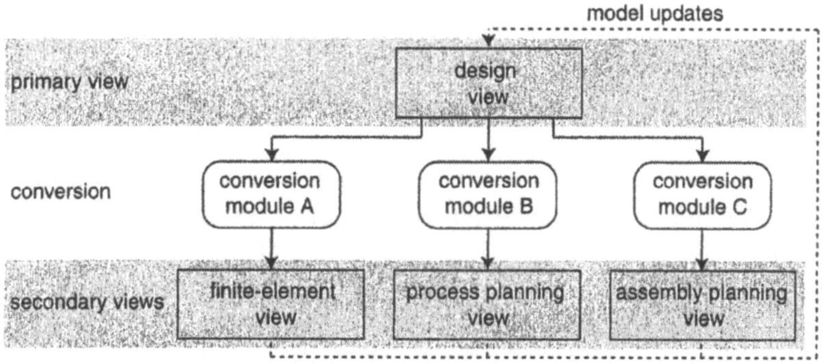

Figure 1. Architecture with primary view and secondary views (after Cunningham and Dixon (1988)).

De Martino et al. (1994, 1998) describe a system for multiple-view feature modelling including one-way feature conversion. The feature models of the views are integrated by an intermediate model that contains, among other things, a representation of the product by means of so-called shape features, i.e. generic protrusions and depressions. The intermediate model is built by a shape-feature recogniser. A new feature model for a view is created from this model by an application-feature converter.

Suh and Wozny (1997) propose a multiple-view feature modelling system with one-way feature conversion using a graph-based intermediate model. The feature model for an application is recognised in the intermediate model by a hybrid feature recognition method that uses a graph-based technique to search features in the intermediate model, given the constraint graph of a feature class, and a hint-based technique to reduce the time that is needed to find a feature.

Bouras et al. (1999) describe a one-way feature conversion approach to create feature models of a product that are suitable for analysis applications. First, a morphological analysis is performed on the shape of the product to create a set of shape features. Subsequently features corresponding to a chosen viewpoint are generated from these shape features. Finally, the feature model generated in this way is simplified and idealised to obtain the so-called working elements, which can be used in an analysis application.

However, support for feature conversion in a single direction only is far from ideal, because it requires an adjustment to a product model to be made in a feature model that may significantly differ from the feature model in which the need for the adjustment occurs. To allow a product model to be adjusted in the feature model of the view in which the need for it occurs, multiple-way feature conversion has been developed.

Bronsvoort and Jansen (1993) propose multiple-way feature conversion as the ideal approach to multiple feature views. Later on, de Kraker et al. (1995, 1997), Bronsvoort et al. (1997), Dohmen (1998) and de Kraker (1998) describe its implementation in a prototype multiple-view feature modelling system. They use a geometric-reasoning-based feature recognition algorithm that derives the feature model of a new view from a non-manifold intermediate cellular representation (Bidarra et al. 1998), and a special kind of coplanar constraints, called link constraints (Dohmen et al. 1996), to propagate changes made in the feature model of one view to the feature models of the other views.

Jha and Gurumoorthy (2000a, 2000b) describe two algorithms that can be used in multiple-way feature conversion. The first algorithm extracts a feature model from a geometric model, based on information whether a particular face of an instance of a feature class creates a new face in the feature model. The second algorithm propagates adjustments in one feature model to the other feature models, based on a special tree data structure that incorporates all feature models for a product.

Hoffmann and Joan-Arinyo (2000) propose an architecture in which the models for different applications are distributed over different CAx systems, and a master model is used to associate the models. This architecture effectively performs multiple-way feature conversion between feature views, at least if the models in these systems are feature models. Each of the systems is said to be a client of the master model server (see *Figure 2*).

The approaches to multiple feature views described above, which all use form features, share a number of shortcomings.

First, all approaches focus on the later product development phases, in which the geometry of the product has to be fully specified. They use the geometry of the product as a basis, and, therefore, cannot be applied in the early product development phases, such as conceptual design.

Second, all approaches deal with a single part only. Real products, however, rarely consist of a single part. Dealing with products that consist of multiple parts, i.e. assemblies, does not only involve the separate parts, but also the relations between these parts.

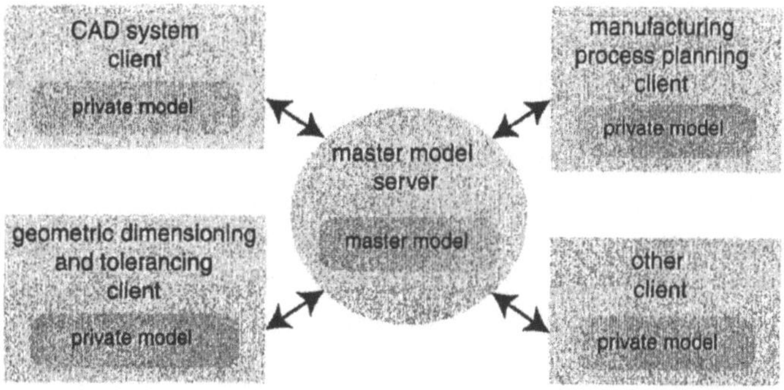

Figure 2. Clients manage their own private model, but can deposit some of their information in the model that is managed by the master model server (after Hoffmann and Joan-Arinyo (2000)).

To support the early product development phases and products with multiple parts, the enhanced multiple-view feature modelling approach has been developed.

3. ENHANCED MULTIPLE-VIEW FEATURE MODELLING

The enhanced multiple-view feature modelling approach can support applications from product development phases in which the geometry of the product has not been fully specified and phases that deal with multiple parts and the relations between them. Views have been developed to support the conceptual design phase, the assembly design phase, the part detail design phase and the part manufacturing planning phase. Some preliminary ideas on enhanced multiple-view feature modelling have been presented in (Noort and Bronsvoort 2000); here the approach is presented in much more detail.

The conceptual design view supports the conceptual design phase. Several approaches to conceptual design exist; three frequently used ones are shortly described. First, in the case-based reasoning approach, a conceptual design is based on a design for a similar product that is chosen from a set of existing designs, and is adapted to fit the requirements for the current product (Maher and Pu 1997). Second, in the sketching approach, a conceptual design is created from sketches that are made by the designer in some drawing application and converted into 3D geometry by the modelling system (Lipson and Shpitalni 2000). Third, in the configuration design approach, a conceptual design is created by specifying the components and the way these are connected in the product (Wielinga and Schreiber 1997). The configuration design approach will be supported here.

The assembly design view supports the assembly design phase. Many approaches to assembly design exist. Some support all phases of product design that involve multiple parts, and provide several tools to analyse, for example, whether the components can be assembled (Zha et al. 2001), others focus on a specific aspect, such as the mate and contact relations between components (Whitney and Mantripragada 1998), or ways to represent complex assemblies (Heissermann and Mattikalli 1998). The assembly design approach that is supported here focuses on the detail design of the connections between the components (van Holland and Bronsvoort 2000).

A part detail design view supports the detail design phase of a part. In this phase, the details of the geometry of the part are specified.

A part manufacturing planning view supports the planning of the way a part can be manufactured, taking into account the capabilities of the available manufacturing equipment.

The views are elaborated in Sections 4 to 6.

Each view has its own feature model of the product, with instances of feature classes from a view-specific library. Each class contains a set of feature constraints, which represent generic requirements from the development phase related to the view. Because only instances of the feature classes from the feature library of a view can be used to build a feature model for that view, each feature model of a view satisfies the generic requirements from the development phase related to that view. In addition to the constraints in feature instances, the feature model can also contain model constraints that represent specific requirements.

The views can be divided into a group of views that deal with all components of a product and the relations between them, i.e. the aggregate-oriented views, and a group of views that deal with a single part only, i.e. the part-oriented views. The conceptual design view and the assembly design view are the aggregate-oriented views, and the part detail design views and the part manufacturing planning views are the part-oriented views (see *Figure 3*). The part-oriented views on a part should represent the same part. The aggregate-oriented views should represent the same product, and represent the parts from the part-oriented views as (sub)components that are related. If the views satisfy these requirements, they are said to be consistent.

Consistency maintenance is elaborated in Section 7.

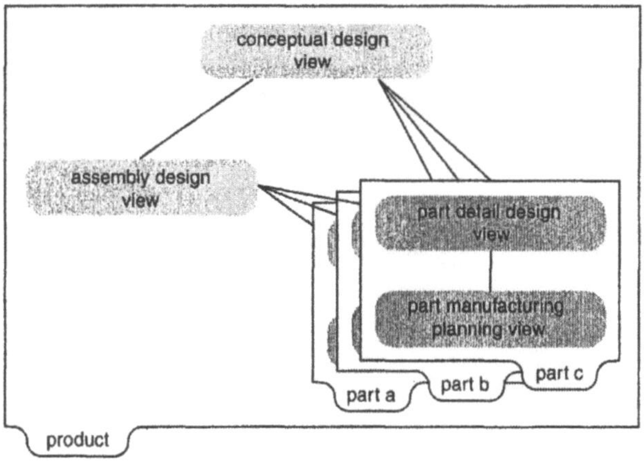

Figure 3. The relations between the part-oriented views and the aggregate-oriented views.

4. CONCEPTUAL DESIGN VIEW

The feature model of the conceptual design view is built from components and interfaces between components. In this way, the conceptual design view can support configuration design. The complete geometry of the components does not have to be specified.

A component in the conceptual design view consists of a base shape with shape concepts and references attached to it, and has attributes such as weight, volume, material type, and a description of its function. Each component fulfils a specific function within the product, which is specified by the designer by means of the shape concepts and the function description.

The base shape is used to position the shape concepts and the references of the component, and gives an impression of the geometry of the component. It is built from simple volumetric shape elements with block, vblock, cylinder or cone shape and one or more constraints to attach the element to the rest of the base shape. In addition, geometric constraints can be used to further position the shape elements with respect to each other.

Shape concepts are more or less abstract entities that represent a specific function of a component and have a shape and a type. The shape can be block, cylinder, cone, vblock or unknown. The type specifies the nature of the shape, i.e. whether it adds material to or removes material from the component, and the way the shape concept is positioned on the base shape or another shape concept. It can be aligned protrusion (Figure 4 (a)), isolated protrusion (*Figure 4*(b)), isolated depression (*Figure 4*(c) top), passage (*Figure 4*(c) bottom), open depression (*Figure 4*(d)), or unknown. The values for the dimensions of the shape and the faces of the other entities in the model that are used to position the shape concept can also be specified, but this is not required. In addition, the function of the shape concept may be specified by means of a description.

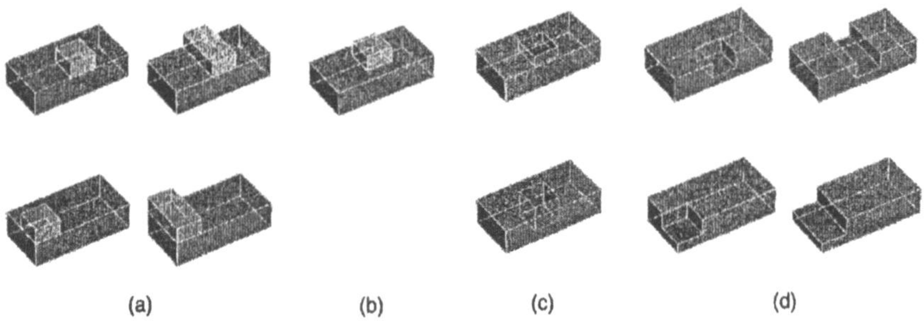

Figure 4. Different types of shape concepts: aligned protrusion (a), isolated protrusion (b), isolated depression and passage (c), and open depressions (d).

References are used to position other model elements, such as shape concepts, and to relate interfaces to the component. They can be points, lines and faces.

Constraints can be used to restrict, for example, the value of the volume attribute of a component or the relative position of two shape concepts, and thus represent functional information. Different types of constraints are available, such as algebraic constraints that restrict attributes of components and shape concepts to be, for example, equal to some value or in some interval, and geometric constraints that restrict the position and orientation of shape concepts or references.

Interfaces between components in the conceptual design view represent the freedom between the components, i.e. how they can move relative to each other. Interfaces are specified with respect to references on components.

Pre-defined types of interfaces, such as hinge, are provided to the designer to be used in the model (see *Figure 5*). By default, interfaces only specify the type of freedom left. Optionally, the direction of the freedom may also be fixed by using the position and orientation of references on which the interface is specified. This approach again allows the designer to exactly specify what is required, but leave irrelevant details unspecified.

To show the feature model to the designer, geometry and relational graph cameras can be used.

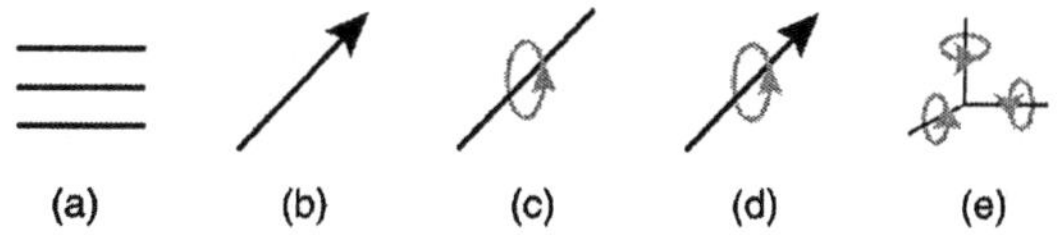

Figure 5. Icons for different types of interfaces: fixed interface (a), prismatic interface (b), hinge interface (c), cylindrical interface (d) and spherical interface (e).

Geometry cameras show the base shape of the components, and can also show the shape concepts on the components and the references (see *Figure 6*(a)). Whether, and how, a shape concept is shown in a geometry camera depends on the amount of information that has been specified for the shape concept; it is not visualised if the type has been specified as unknown, its geometry is visualised if type, shape and all other parameters have been specified, and it is visualised by an icon based on the type and/or shape otherwise. Whether references are shown in a geometry camera, depends on the fact whether these have been selected for visualisation by the designer or not.

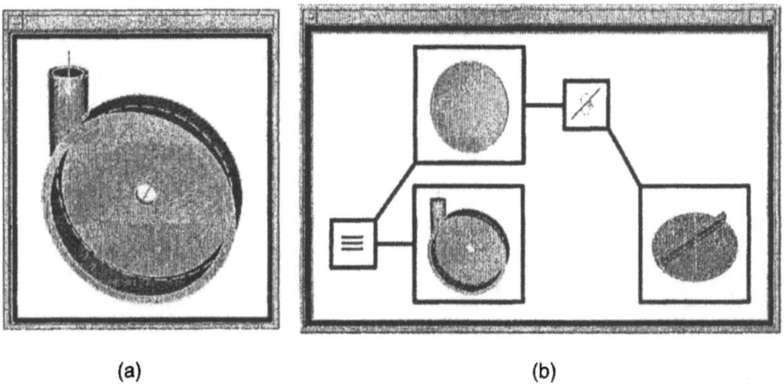

Figure 6. A geometry camera (a) and a graph camera (b) on the feature model of the conceptual design of a water pump.

Graph cameras show the components and the interfaces between them. A component is represented by a node that contains an image of the component. An interface is represented by a node that contains an icon for the interface (see *Figure 5*), and is connected to the nodes that represent the components that are related by the interface. An example of a graph camera is shown in *Figure 6*(b).

The model operations that can be performed on the feature model of the conceptual design view are adding, removing and editing components and interfaces. Editing components, for example, may involve changing the base shape, the functional description, and adding, removing or editing shape concepts, references or constraints on the component. The values for the parameters of these model operations can be specified via a modelling panel or via selection in a camera.

An example of the use of the conceptual design view for the development of a modern version of the historical high-wheel bicycle (*Figure* 7(a)) is given here. The model of the conceptual design view represents the components of the product, with their important aspects only, and the interfaces between the components. *Figure* 7(b+c) show some cameras on the model.

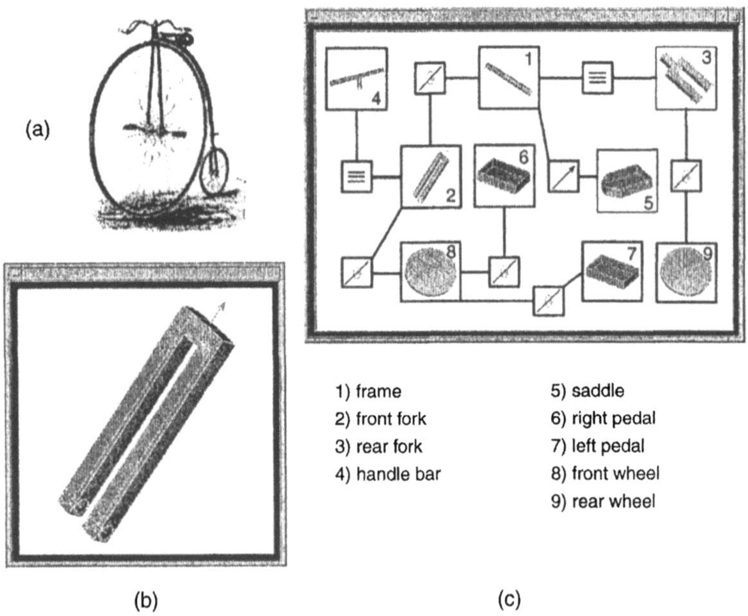

Figure 7. The historical high-wheel bicycle (a), and some cameras on the model of a modern version of it: the base and the shape concepts of the front fork component in a geometry camera (b), and the interfaces between the components in a graph camera (c).

5. ASSEMBLY DESIGN VIEW

The feature model of the assembly design view consists of components and connection features between them, which together form assemblies. The assembly feature model can contain multiple unconnected assemblies and components, but no unconnected connection features.

Components are the elements of the assembly feature model that are combined, and on which the assembly information is defined. A component is centred around a reference frame, i.e. all elements of a component are (in)directly specified relative to this frame. The components have the form features of the connection features that have been specified on them attached to them (see below), but no other form features.

A component is either a single component, which is based on a part, or a compound component, which is based on an assembly. A single component represents a part in the assembly model. A compound component encapsulates an assembly for further assembly modelling purposes, by hiding its internal structure of components and connection features, and dealing with the boundary of the assembly only.

Connection features are functional relations that represent assembly information, such as the internal freedom of the connection, the types of the form features needed for the connection on the components, and the way the connection can be established (van Holland and Bronsvoort 2000). Constraints are used to specify part of this assembly information. For example, in a dove-tail connection feature, contact constraints are used between faces of the dove-tail rib form feature and the dove-tail slot form feature to specify that the rib should stay in the slot. Other examples of connection features are a rib-slot and a pin-hole connection feature.

To show the feature model of the assembly design view, geometry cameras, relational graph cameras and hierarchical graph cameras may be created (Bronsvoort et al. 2002).

The geometry that is displayed in a geometry camera on a component consists of the so-called reference geometry, i.e. the geometry of the part or assembly related to that component, and in addition the geometry of the form features of the connection features on that component. The reference geometry is included to give a better insight into the complete geometry of the component, but is visualised without shading to emphasise that it is not part of the feature model for the component in the assembly design view. It is also used to enable a more intuitive way of creating form features of the connection features on the component: these need not to be specified with respect to the reference frame or other form features, but can also be specified with respect to the faces of the reference geometry.

Relational graph cameras display the connections between the components in an assembly (see *Figure 8*(a)).

Hierarchical graph cameras display the hierarchy of an assembly with its components (see *Figure 8*(b)), and optionally subcomponents of compound components (see *Figure 8*(c)). They use nodes containing an image of the component to represent components, and represent the hierarchy between compound components and their subcomponents by edges between their nodes.

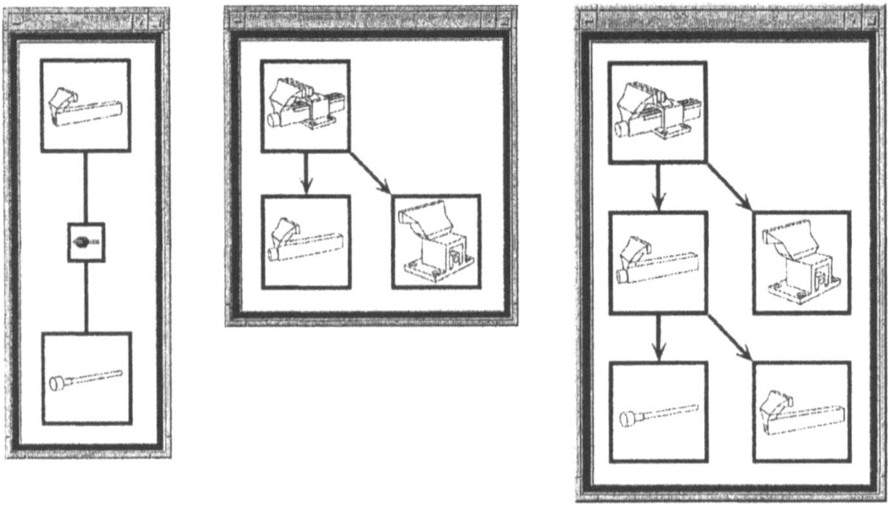

(a) (b) (c)

Figure 8. Graph cameras showing a relational graph of an assembly (a), a hierarchical graph of the assembly (b), and the hierarchical graph including the subcomponents of one compound component (c).

The model operations that can be performed on the feature model of the assembly design view are creating and removing components, adding a connection feature between components, editing the parameters of a connection feature, and removing a connection feature. The values for the parameters of these model operations can be specified via a modelling panel or via selection in a geometry or graph camera.

Adding a connection feature between components requires the appropriate form features, e.g. a dove-tail rib and a dove-tail slot for a dove-tail connection, on these components. If these form features do not yet exist in the assembly design view, they have to be created. This results in the need for the feature models of the other views to be updated in order to keep all views consistent (see Section 7). Removing a connection feature also removes the related form features, at least if they are not used by any other connection feature: form features only exist in the assembly design view if they are related to a connection feature, otherwise they are removed.

An example of the use of the assembly design view for the high-wheel bicycle from the previous section is given here. The example shows the connection of the rear wheel to the rear fork with an additional axis component that is connected to the rear fork with a fixed pen-hole connection and to the wheel using a rotating pen-hole connection (see *Figure 9*).

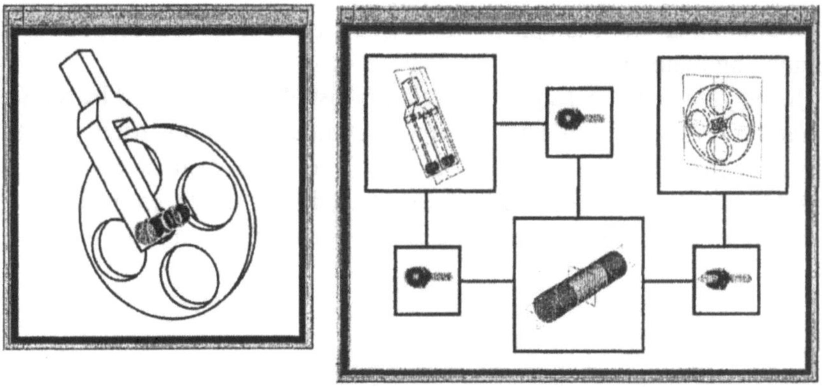

Figure 9. The rear wheel of the bicycle is connected to the rear fork using an axis component.

6. PART DETAIL DESIGN AND MANUFACTURING PLANNING VIEWS

The form feature models of a part detail design view and a part manufacturing planning view are built from form features and additional model constraints (Bidarra and Bronsvoort 2000).

In a part detail design view or manufacturing planning view, the form features contain class-specific design information that is captured by means of feature elements and feature constraints. Feature elements are shapes and user-defined variables. Feature constraints can be, for example, a geometric distance face-face constraint, a dimension constraint, which specifies a dimension to be within a given range, and an on-boundary constraint, which specifies a feature face to be on the boundary of the part. Examples of form features are block protrusion and rounded blind slot.

Model constraints are used to specify additional functional information that is related to only one feature instance, and functional information that is related to several feature instances. They can be of the same types as the feature constraints.

To show the shape of the form feature model of a part detail design view or a manufacturing planning view, geometry cameras are used that combine several visualisation techniques (Bronsvoort et al. 2002). These cameras provide functional insight into the feature model by visualising all sorts of engineering information. This includes all features of a specific class to show all regions of the geometry with a similar function, intersections between features to show all regions of the geometry that have been specified more than once, and information on the fact whether a feature face is on the boundary of the part or not to show all regions that can be formed by removing material.

The way this engineering information is visualised is defined by the user: several line visualisation and shading techniques may be used for it, in various combinations. An example of a geometry camera showing the feature faces of some selected features shaded, and the faces of the rest of the model with dashed-hidden-line visualisation, is given in *Figure 10.*

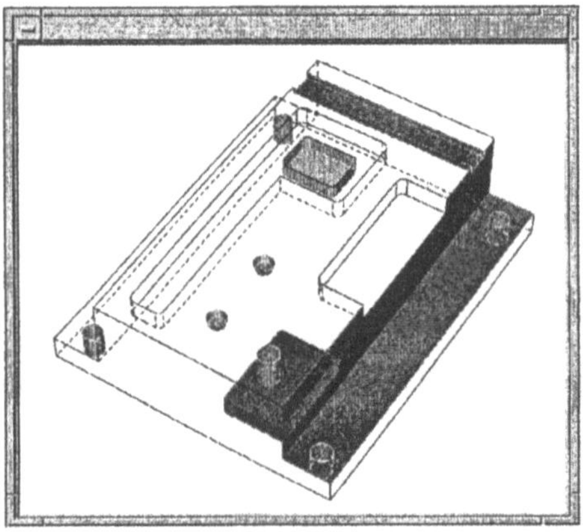

Figure 10. A geometry camera that combines several visualisation techniques to provide functional insight into a feature model.

The model operations that can be performed on the feature model of the view are adding a new form feature or model constraint, editing the parameters of an existing feature or constraint, and removing an existing feature or constraint.

After each operation, the validity of the model is checked. The model is no longer valid if one or more constraints are not satisfied. If an invalid situation is found, the operation is held up until the invalid situation has been removed by the user by additional, corrective model operations (Bidarra and Bronsvoort 2000).

The values for the parameters of these model operations can be specified via a modelling panel or via selection in a geometry camera.

An example of the use of the part detail design and manufacturing planning view for a part of the high-wheel bicycle of Section 4 is given here. The example shows the detail geometry of the part that is used to fix the saddle to the frame, and the model of the manufacturing planning view of this part that can be used to determine the way the part can be manufactured (see *Figure 11*).

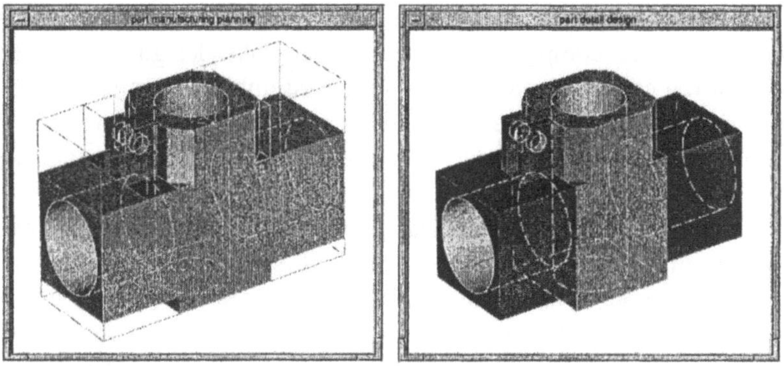

Figure 11. The model of the detail design view of the part that is used to fix the saddle (a), and the model of the part manufacturing planning view that is derived from this model (b).

7. CONSISTENCY MAINTENANCE

Consistency maintenance is used to update the feature model of a view in order to make it consistent with the feature model of another view. Because of the different natures of the views in the enhanced multiple-view feature modelling system, the feature model of a view cannot always be derived from the feature model of an arbitrary other view, and different techniques are needed for different pairs of views.

To support the use of the views described in the previous sections, consistency maintenance needs to be performed only for the source-target pairs that are indicated by an edge in *Figure 3.*

These pairs of views are a minimal set of pairs of views to maintain the consistency of the feature models of all views. This is important because the number of consistency maintenance steps that need to be performed should be minimised to reduce the number of consistency maintenance techniques required, and to optimise the performance.

Model elements in different views that should be kept consistent, can be related by a designer by specifying associations between them. For example, associations can be specified between a component in the conceptual design view and a component in the assembly design view, between a component in the conceptual design view and one or more parts in their detail design view, and between a shape concept of a conceptual component and a form feature of an associated assembly component or part.

Consistency maintenance between the conceptual design view and the assembly design view consists of keeping the components in the conceptual design view consistent with the associated components in the assembly design view, and keeping the interfaces in the conceptual design view consistent with the associated connections in the assembly design view. Notice that some components in the conceptual design view may be associated to a compound component in the assembly design view, and as a result some (single) components in the assembly design view are only via a compound component associated to a component in the conceptual design view.

A component in the conceptual design view is consistent with a component in the assembly design view if each shape concept of the conceptual component that is associated to a form feature of the assembly component is consistent with that form feature. Notice that a shape concept in the conceptual design view does not have to be related to a form feature in the assembly design view, and vice-versa.

A shape concept is consistent with a form feature if the shape, type and dimensions of the form feature satisfy the constraints that have been specified on the corresponding attributes of the shape concept.

If a shape concept is inconsistent with the associated form feature, the designer has to adjust the shape concept or the form feature, or to remove the association.

For each interface in the conceptual design view, a connection feature in the assembly design view has to exist. An interface is consistent with a connection feature if the types of freedom that are reduced by the interface and the connection feature are equal, and the direction of this reduced freedom in the interface, if it has been specified, corresponds to the direction in the connection feature.

If an interface is inconsistent with the associated connection feature, the designer has to adjust the interface or the connection feature, or to remove the association.

Consistency maintenance between the conceptual design view and the part detail design views consists of keeping the components in the conceptual design view consistent with the associated parts in their detail design view. This involves keeping the shape concepts of the components consistent with the associated form features (see above), and keeping the attributes of the components consistent with the physical properties of the associated parts. Notice that the feature models of the detail design views on parts may contain additional form features that are not related to any shape concept.

The attributes of a component are consistent with the physical properties of the associated parts if, for example, the sum of the values of the volume attributes of the parts is within the range prescribed for the volume attribute of the component.

If the attributes of the component are inconsistent with the associated parts, the designer has to adjust the parts, or change the requirements on the component, e.g. the range for a value attribute.

Consistency maintenance between the assembly design view and the part detail design views consists of keeping each single component in the feature model of the assembly design view consistent with the feature model of the detail design view on the associated part.

This is the case if the faces of the form on the component that are intended to be on the boundary of the component, are on the boundary of the feature model of the associated part. Notice that faces of form features in the feature model of the detail design view on a part that are on the boundary of the part do not have to exist among the faces of the features in the feature model of the assembly design view.

Automatic feature recognition is used to make the feature model of a part in a detail design view that is inconsistent with its associated component in the assembly design view consistent again (de Kraker et al. 1995, 1997, Bronsvoort et al. 1997). Feature linking is used to keep the feature models of the parts in the detail design views consistent with the associated components in the assembly design view during simple adjustment of the feature model of either. The system automatically creates links between the faces of the features from different views that are co-planar and on the boundary of the part or the component. A link consists of a geometric co-planar constraint (Dohmen et al. 1996).

Consistency maintenance between the detail design view and the manufacturing planning view of a part consists of keeping the feature model of the part detail design view consistent with the feature model of the part manufacturing view.

This is the case if they represent the same geometry.

Again automatic feature recognition is used to make the feature model of a part detail design view that is inconsistent with the feature model of the manufacturing planning view on the same part consistent again, and vice versa. Similarly, feature linking is again used to keep the feature model of the part detail design view consistent with the feature model of the part manufacturing planning view during simple adjustment of the feature model of either (see above).

An example of consistency maintenance of the feature models of different views on the high-wheel bicycle of Section 4 is given here. The example shows that the feature models of two part detail design views and the assembly design view are kept consistent after a change in the feature model of one detail design view; see *Figure 12*.

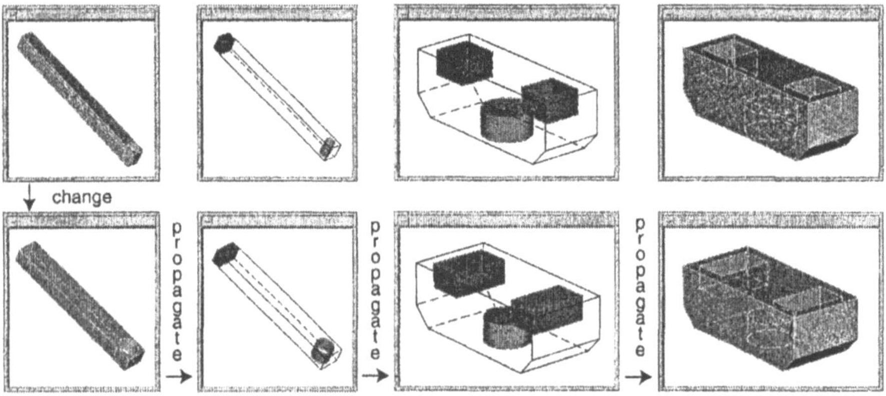

Figure 12. A change in the detail design view of a leg of the front fork, is propagated to the assembly design view of the associated component, and subsequently to the assembly design view of the connected component and the detail design view of the associated part.

8. CONCLUSIONS

This paper presented the enhanced multiple-view feature modelling approach, which is able to support a wider range of product development phases than previous multiple-view feature modelling approaches. In contrast to these approaches, the new approach also supports development phases in which the geometry of the product has not been fully specified, and product development phases that deal with multiple parts and relations between them.

Further, it allows the feature model of any view to be adjusted, while keeping all views consistent. The techniques that are used for this have been described.

A prototype system has been implemented with a conceptual design view, an assembly design view, part detail design views and part manufacturing planning views.

The feature models of the various views, which represent only those aspects of a product that are relevant in the associated product development phase, have been found to considerably increase the insight of the designer into the product model. The designer is not disturbed by aspects of the model that are irrelevant during that phase, but can concentrate on the aspects that are relevant.

Altogether, it can be concluded that the enhanced multiple-view feature modelling approach is capable of supporting a wider range of product development phases, and thus to better support product development.

REFERENCES

Bidarra, R. and Bronsvoort, W. F. (2000). Semantic feature modelling, Computer-Aided Design, Vol. 32, No. 3, pp. 201-225.

Bidarra, R., de Kraker, K. J. and Bronsvoort, W. F. (1998). Representation and management of feature information in a cellular model, Computer-Aided Design, Vol. 30, No. 4, pp. 301-313.

Bouras, A., Belaziz, M., Tehari, A., Wahu, G. and Brun, J. M. (1999). A strategy to support application's dependent features interpretation, Journal of Intelligent Manufacturing, Vol. 10, No. 1, 41-47.

Bronsvoort, W. F., Bidarra, R., Dohmen, M., van Holland, W. and de Kraker, K. J. (1997). Multiple-view feature modelling and conversion, In Strasser, W., Klein, R. and Rau, R. (eds.): Geometric Modeling: Theory and Practice - The State of the Art, pp. 159-174, Springer-Verlag.

Bronsvoort, W. F., Bidarra, R. and Noort, A. (2002). Feature model visualization, To be published in: Computer Graphics Forum, Vol. 21, No. 4.

Bronsvoort, W. F. and Jansen, F. W. (1993). Feature modelling and conversion - Key concepts to concurrent engineering, Computers in Industry, Vol. 21, No. 1, pp. 61-86.

Bullinger, H. J. and Warschat, J. (1995). Concurrent Simultaneous Engineering Systems, Springer-Verlag.

Cunningham, J. J. and Dixon, J. R. (1988). Designing with features: the origin of features, In Tipnis, V. A. and Patton, E. M. (eds.): Proceedings of the 1988 ASME International Computers in Engineering Conference and Exhibition, Vol. 1, pp. 237-243.

De Martino, T., Falcidieno, B., Giannini, F., Haβinger, S. and Ovtcharova, J. (1994). Feature-based modelling by integrating design and recognition approaches, Computer-Aided Design, Vol. 26, No. 8, pp. 646-653.

De Martino, T., Falcidieno, B. and Haβinger, S. (1998). Design and engineering process integration through a multiple view intermediate modeller in a distributed object-oriented system environment, Computer-Aided Design, Vol. 30, No. 6, pp. 437-452.

Dohmen, M. (1998). Constraint-based Feature Validation, PhD Thesis, Delft University of Technology.

Dohmen, M., de Kraker, K. J. and Bronsvoort, W. F. (1996). Feature validation in a multiple-view modeling system, In McCarthy, J. (ed.): CD-ROM Proceedings of the 1996 ASME Design Engineering Technical Conferences and Computers in Engineering Conference.

Heissermann, J. and Mattikalli, R. (1998). Representing relationships in hierarchical assemblies, In CD-ROM Proceedings of the 1998 ASME Design Engineering Technical Conferences and Computers in Engineering Conference.

Hoffmann, C. M. and Joan-Arinyo, R. (2000). Distributed maintenance of multiple product views, Computer-Aided Design, Vol. 32, No. 7, pp. 421-431.

van Holland, W. and Bronsvoort, W.F. (2000). Assembly features in modeling and planning, Robotics and Computer Integrated Manufacturing, Vol. 16, No. 4, pp. 277-294.

Huang, G. Q. (1996). Design for X, Chapman and Hall.

Jha, K. and Gurumoorthy, B. (2000a). Multiple feature interpretations across domains, Computers in Industry, Vol. 42, No. 1, pp. 13-32.

Jha, K. and Gurumoorthy, B. (2000b). Automatic propagation of feature modification across domains, Computer-Aided Design, Vol. 32, No. 12, pp. 691-706.

de Kraker, K. J. (1998). Feature Conversion for Concurrent Engineering, PhD Thesis, Delft University of Technology.

de Kraker, K. J., Dohmen, M. and Bronsvoort, W. F. (1995). Multiple-way feature conversion to support concurrent engineering, In Hoffmann, C. M. and Rossignac, J. (eds.): Solid Modeling '95, Third Symposium on Solid Modeling and Applications, pp. 105-114.

de Kraker, K. J., Dohmen, M. and Bronsvoort, W. F. (1997). Maintaining multiple views in feature modeling, In Hoffmann, C. M. and Bronsvoort, W. F. (eds.): Solid Modeling '97, Fourth Symposium on Solid Modeling and Applications, pp. 123-130.

Lipson, H. and Shpitalni, M. (2000). Conceptual design and analysis by sketching, AI-EDAM - Artificial Intelligence for Engineering Design, Analysis and Manufacturing, Vol. 14, No. 5, pp. 391-401.

Maher, M. and Pu, P. (1997). Issues and Applications in Case-Based Reasoning in Design, Lawrence Erlbaum Associates.

Noort, A. and Bronsvoort, W. F. (2000). Enhanced multiple-view feature modelling, In Brunet, P., Hoffmann, C. and Roller, D. (eds.): CAD Tools and Algorithms for Product Design, pp. 80-94, Springer-Verlag.

Shah, J. J. and Mäntylä, M. (1995). Parametric and Feature-based CAD/CAM, John Wiley and Sons, Inc.

Suh, Y. S. and Wozny, M. J. (1997). Interactive feature extraction for a form feature conversion system, In Hoffmann, C. M. and Bronsvoort W. F. (eds.): Solid Modeling '97, Fourth Symposium on Solid Modeling and Applications, pp. 111-122.

Whitney, D. E. and Mantripragada, R. (1998). The Datum Flow Chain: A systematic approach to assembly design and modeling, In CD-ROM Proceedings of the 1998 ASME Design Engineering Technical Conferences and Computers in Engineering Conference.

Wielinga, B. and Schreiber, G. (1997). Configuration-design problem solving, IEEE Expert, Vol. 12, No. 2, pp. 49-56.

Zha, X. F., Du, H. J. and Qiu, J. H. (2001). Knowledge-based approach and system for assembly oriented design, part i: the approach, Engineering Applications of Artificial Intelligence, Vol. 14, No. 1, pp. 61-75.

Improving Design Productivity and Product Data Consistency

From Design Intent to Solid Models through Technological Links

F. GIGUÈRE[1], L. RIVEST[2] and A. DESROCHERS[1]
1 University of Sherbrooke, Department of Mechanical Engineering;
2) École de technologie supérieure, Department of Automated Production Engineering

Abstract: Computer Aided Design systems still suffer from a number of shortcomings with regards to the full design process: they provide limited solutions for maintaining product data consistency and offer no assistance to relieve designers from tedious, repetitive, solid modeling work. This paper presents a framework for improved support of the design process from preliminary to detailed design phase. The proposed solution aims at managing product data consistency within the assembly model as well as at assisting designers with specific geometric definition tasks by exploiting the design context. First, relationships between parts are captured by establishing specialized technological links, called *derivation links*, that encapsulate specific design knowledge. These links are then used to apply design knowledge to the design context and to derive design feature geometry, called *pseudo-imprints*. These links are maintained and later exploited to control design change propagation. An aircraft structure design example illustrates these concepts.

Key words: Features, Assembly Modeling, Feature-Based Design, CAD, Modeling Method

1. INTRODUCTION

Products rarely consist of a single part. They are usually composed of an array of assembled parts, allowing different materials to be used. Generally, parts of different sizes and shapes are best produced separately. The use of computers in design has allowed modelling these parts as 3D solids and to assemble them virtually before actual manufacture. Computer aided design (CAD) tools have therefore helped reduce the number of errors before the manufacturing stage as well as the number of prototypes needed. While they have many advantages, these methods still present several problems. One of these is that modelling complex parts is time consuming even for highly skilled designers, who are required to carry out detailed and tedious tasks. Another problem is the build-up of massive databases containing inconsistent information, as a result of the increasing capability to create multiple versions of electronic documents and the availability of cheap data storage.

With regard to the first problem, the use of feature-based design has improved the situation by allowing designers to work at higher levels of abstraction. Portions of geometry are instanced, sized and located quickly by specifying a minimum set of parameters while the feature modeller completes the details. This modelling method is starting to emerge in commercial CAD software, in which, however, it is usually applied to the design of parts considered individually rather than within the design context.

In terms of the second problem, little has been done so far. Assemblies are typically modelled as a hierarchy of subassemblies and components, but current CAD software captures little of the semantics of relationships between parts. Currently, propagating a modification made to one part of an assembly requires the designer to manually perform a series of tedious operations to update all impacted models accordingly, in order to maintain the global coherence of the assembly model. Relying on such manual interventions easily leads to design errors that are sometimes revealed on the shop floor. The use of assembly features in assembly modelling is a topic of research aiming to capture relationships between parts in order to solve this data consistency problem.

A four-year research project involving Bombardier Aerospace, Université de Sherbrooke and École de Technologie Supérieure was undertaken in 1998 to define an information system architecture that would allow exploiting CAD data with Product Data Management tools. In this paper, we focus on the two problems mentioned above (data consistency and tedious product modelling tasks). A framework is proposed that manages multi-model links to ensure information consistency while increasing the productivity of specific modelling tasks. It allows the capture of design intent by establishing technological links between features belonging to different parts. These links are specialized so as to embed knowledge relevant to specific design tasks. Specialized applications are triggered to assist designers in creating a feature "pseudo-imprint" geometry obtained from a "derivation link".

2. RELATED WORK

In this section, research works falling into the areas of Design processes, Features and Assembly modelling, are briefly reviewed. In the area of design processes, the bottom-up approach has been perceived to be the most widely used for assembly modelling within CAD systems. This approach requires that parts be fully defined before creating the assembly. However, a more natural way is to model assemblies with the top-down approach, which Libardi (Libardi 88) defines as a recursive subdivision of an initial abstract concept into logical sub-assemblies until the part level is reached. This approach allows to better support the design process where the designer thinks in terms of abstract function before being concerned with the geometry (Ullman 97).

Research on assembly modelling has been conducted in various ways with different goals in mind. Several assembly models have been proposed (Rimscha 90) (Sodhi 91) (Mailharro 98) (Bronsvoort 94). Gui and Mäntylä (Gui 93) distinguish themselves by proposing a generic assembly model capable of supporting both conceptual design and detailed design, developing a design case-based reasoning system. They use an assembly-modelling scheme with a function-based approach, "first assembly then components". Therefore they use a top-down process which is not limited to merely designing with features and which is closer to the way designers work. Their proposal aims at taking into account the assembly at the early stage of the design process. In a following paper (Gui 94) they develop these concepts further by presenting a general assembly model that provides a function-oriented view for design knowledge and a module-oriented view for manufacturing knowledge. They state that features are too low-level to support the early phases of design.

Shah and Mäntylä (Shah 95) define a form feature as the generic shape or characteristic of a product with which engineers can associate attributes and knowledge for reasoning about the product. The concept of assembly feature is defined as an association between two form features which belong to different parts. Different approaches to assembly modeling are described and the assembly feature concept is presented with its various advantages:

- Assembly features allow assembly creation at a higher level by storing all of the constraints between the mating features as a single unit.
- Assembly features allow constraints to be expressed in terms of feature dimensions rather than face relations.
- Assembly features allow design changes made to a part to propagate to other parts in an assembly.

Shah (Shah 92) introduces the notion of feature-based assembly modelling. He developed a system for assembly design which encapsulates different constraints between mating features. His research targets bottom-up assembly modelling in the detail design phase. In a subsequent article (Shah 1993), he explores the advantages and limitations of procedural and declarative approaches for assembly modelling. He explains how some structures can be used to define relationships between assemblies, parts, features, and feature volume primitive. He proposes methods for deriving the location of an assembly unit from high-level constraint specifications.

The goal of the FEAST project (Feature-based Assembly Techniques) (Bordegoni 99) (Deneux 99) was to find a way to extend the use of features from the design area to help solve assembly problems. Bordegoni and Cugini (Bordegoni 99) developed concepts on how assembly features can be used to solve assembly problems, and presented a graphical formalism for representing assembly models and structures. They also introduced a process for knowledge formalization focusing on user/system activity sharing and knowledge sharing. An assembly feature-based demonstrator was created to validate these concepts.

Within the same framework Deneux (Deneux 99) advocates assembly features, explaining how they can be characterized in any field concerned with assembly design. He illustrates the assembly structure with a matrix where the relationships between different part features can be emphasized. He also proposes a process for formalizing feature data and behaviour. Finally he presents a demonstrator and a prototype of an assembly modeller for the aeronautic industry.

The work described in this paper differs from the above studies in several aspects. Firstly, while the bottom-up approach is the most widely used in the application of assembly features to the detail design phase of assembly modelling (i.e. assembly modelling using fully defined parts), we propose a method closer to the top-down approach. We begin at the preliminary design phase, where parts are only partially defined. These partially defined parts are placed in the design context and we form technological links to assist in the creation of the missing features. As a result, our method captures the design intent right from the preliminary design phase and exploits it well into the detail design phase. Secondly, our method can be used with any CAD system, rather than with design-by-feature modellers, since the technological links we create between parts do not involve CAD-generated features. Furthermore, we use the technological links (defined further) to capture and reuse design intent and to assist in the creation of certain features. To be useful, these technological links must be strongly typified, meaning that a specific link must be defined for the resolution of a specific design situation. Implementing these concepts of course leads to the development of "task specific" applications providing relevant technological links for the problem at hand, thus endorsing the underlying idea that improving design productivity requires creating specialized CAD tools to replace or complement the generic ones currently available on the CAD market. At the time of writing, these concepts are being implemented in an application to illustrate them. We insist, however, that the proposed concepts remain generic in their nature and that they can be applied to a vast collection of design problems. For example, Macabies and other members of our research team used technological links to create an application that helps design fasteners between parts (Macabies 00).

3. STATEMENT OF WORK

It was evaluated that 80% of the time spent on some detail design tasks does not involve creative work (educated guess based on observation and interviews with designers). Consequently, the main objective of this article is to propose a solution for capturing and reusing knowledge in order to maintain data consistency within the assembly model and to improve, assist and automate repetitive and tedious modeling tasks. To be more specific, we wish to:

- Propose a solution to capture design intent right from the preliminary design phase.
- Use the top-down approach by working with assemblies right from the preliminary design phase when parts are not fully defined.
- Exploit the design context (i.e. the virtual environment of the part) and some design rules to assist the designer in the definition of specific form features.
- Ensure information consistency within the digital mock-up.
- Provide assistance to well-defined and repetitive design tasks.
- Develop applications that are not linked to specific CAD systems and that do not require one to work in a feature-based design environment.

4. CONCEPTUAL ELEMENTS OF THE APPROACH

This project proposes a framework for developing applications on top of current CAD systems in order to assist designers in defining missing features on parts by capturing and exploiting the design intent. These missing features are those whose geometry is strongly constrained by the design context. Hence, when designing a complex product, parts are defined up to a certain point in the design context. They are then linked to other parts of the assembly so that the missing features, whose design depends on the context, may be created with the help of the application. The goal of this approach is first to improve part modelling by facilitating the creation of certain features. Secondly, product data consistency is managed by creating links between related parts so that modifications, such as dimensional changes, can be propagated even after part models have been digitally assembled.

This requires that the application be able to:

- Capture the relationships between related parts by creating a link between them;
- Create a feature using the information derived from the part by the link;
- Preserve this link to allow design change propagation;

To fulfil such requirements, the following concepts will be used: design feature, design context, technological link, derivation link, pseudo-imprint, target features and parts, datum features and parts, assisted creation.

- Design feature: Functional element of an object which has an engineering meaning and that can act as a handle over a set of information (henceforth, we will use the word "feature" to refer to "design feature").
- Design context: The virtual environment in which the designer is working, defined by a portion of the digital mock-up.
- Technological link: A link between two entities that carries technological information and that is strongly typified.
- Derivation link : Specialized technological link that encapsulates a unidirectional relation between a target feature and a datum feature of the design context.
- Pseudo-imprint : Geometry of a target feature on a part determined automatically by applying knowledge to the geometry of another part of the design context by means of a "derivation link".
- Target feature : Instance of a design feature for which a pseudo-imprint is defined.
- Datum feature : Design feature from which the target feature definition is derived.
- Target part : Part on which a target feature is defined and created.
- Datum part : Part to which the datum feature belongs.
- Assisted creation : Automatic construction process over which the user maintains a certain level of control.

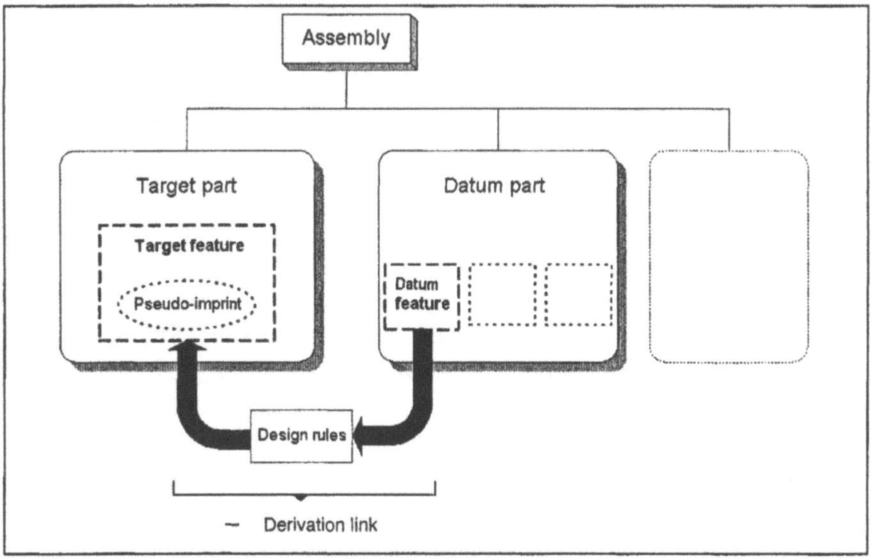

Figure 1. Pseudo-imprint and Derivation link structure

Figure 1 shows a schematic representation of the pseudo-imprint and derivation link concepts. While the "derivation link" may be likened to an assembly feature, unlike the latter it does not relate two pre-existing features. Rather, it relates a datum feature to another part, with the purpose of creating a new feature on the target part, for which the geometrical parameters define a pseudo-imprint. Therefore, the geometry of the derived target feature is a pseudo-imprint of a reference feature on the datum part. This derivation link implies a unidirectional relation. This means that a modification made to the datum part will be propagated to the pseudo-imprint when the geometry of the datum feature changes or when the relative position of both parts is modified.

Generic properties of the target feature are stored in the application. They record the feature's engineering significance and generic shape and provide the basis of its behaviour. These properties allow the shape to be generated by means of pre-defined procedures arising from a set of design rules relevant to a specific problem. Focusing on specific well-defined problems allows us to strongly categorize the derivation link, which in turn improves efficiency by letting us know exactly what we are working with.

The design rules are applied to the independent parameters obtained from the other models in the assembly, i.e., the design context. The technological link established between the datum and the target is used to apply knowledge to these independent parameters and thus derive the geometry of the target feature. This is why we say that a pseudo-imprint is the geometry of a target feature defined by the design context.

5. PROPOSED PROCESS

As mentioned previously, this project seeks to define general concepts to help improve the productivity of some design tasks while ensuring product data consistency. One of the keys to achieving this goal is to focus on specific design problems that can be solved automatically by applying engineering knowledge through dedicated applications. A task-specific oriented method is thus used to restrict the application domain. In order to implement this method in a design methodology, it is necessary to fully characterize the problem at hand by identifying its specific geometric configuration, the criteria of applicability, the specific design rules involved as well as the design and manufacture constraints.

This type of specialization calls for the use of the procedural approach over the declarative approach. In the procedural approach, generic features are pre-defined in terms of rules and procedures, whereas in the declarative approach, the feature definition is coded in a non-procedural language where a collection of specified constraints are resolved. This makes the procedural approach a better choice for handling specific cases, the declarative approach being suitable for generalization (Shah 93). Aside from being easier to implement, the procedural approach is closer to the structure of the design process. Moreover, it avoids conflicts and allows the use of unidirectional links. The derivation link that we introduce, being unidirectional in nature, can therefore be managed through the procedural approach. However, its applicability is restricted to a specific class of design problems.

The proposed steps are shown in *Figure 2*. It begins in Step 1 with the identification of the operational context by the user. In Step 2 the user partially models the part which will become the target and positions it in the design context by means of the digital mock-up tools provided by the CAD system. In Step 3 the user creates the derivation link by identifying both the datum and target parts in the design context. During Step 4 the application establishes the relative position and orientation of the datum in the target coordinate system. In Step 5 the application reads from the datum model all the information needed to create the pseudo-imprint. In Step 6 the application derives the pseudo-imprint dependent parameters by applying the design rules that were implemented. Step 7 creates the pseudo-imprint on the target part. Finally, Step 8 maintains the link between the target and datum parts. Whenever a change is made to the datum part, change propagation is achieved by repeating Steps 4 to 7 for each pseudo-imprint involved.

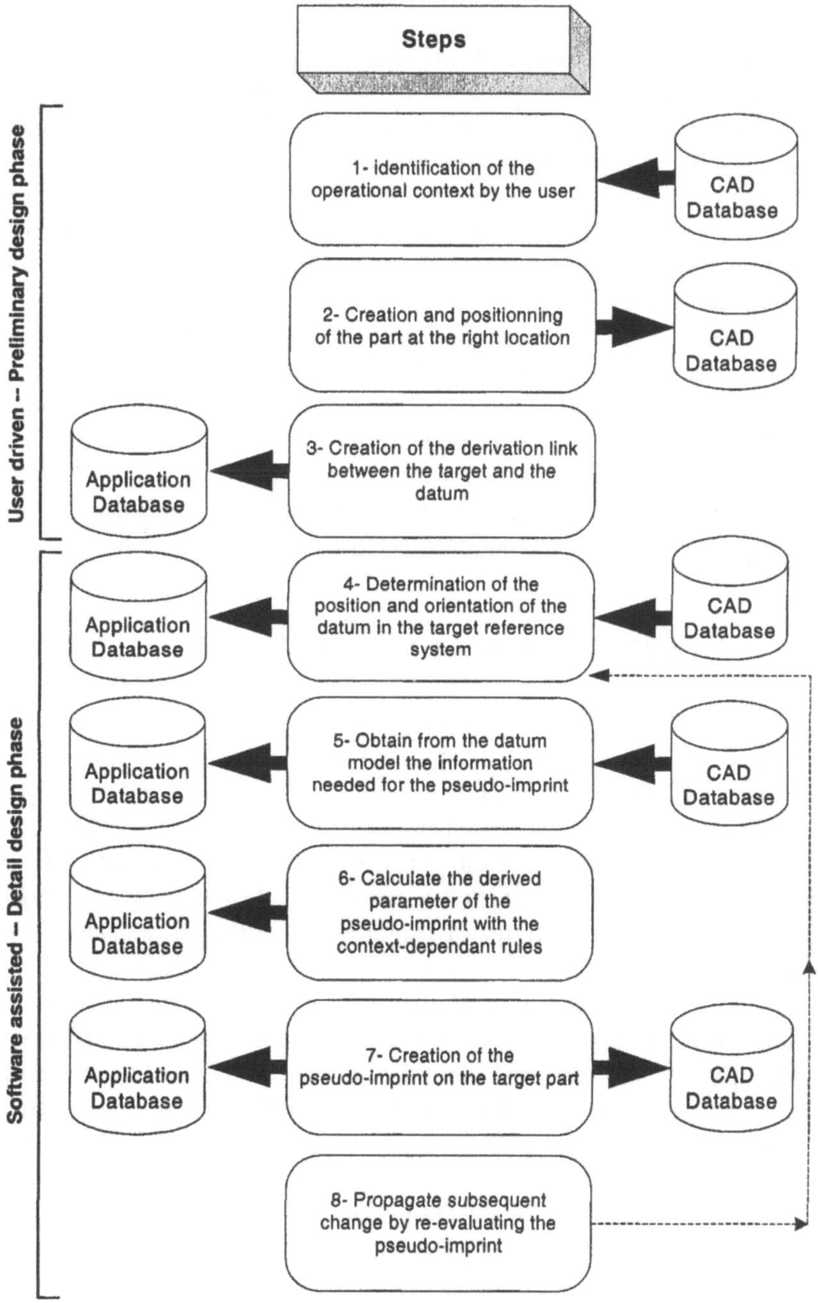

Figure 2. Procedure for the creation of the pseudo-imprint with the derivation link

The CAD database and application database are shown in *Figure 2* to highlight the fact that the application is independent from the CAD platform used. Moreover, because the derivation link is created between features defined by the application rather than by a design-by-feature modeller. Furthermore, *Figure 2* shows that this method fits in the design process by capturing, at the preliminary design phase, the information through user interaction which is then used at the detail design phase in the software assisted creation process.

The proposed approach can be used to solve design problems where the geometry of a portion of a part can be derived from the design context and where the specialized technological link can be exploited often enough to justify investing in such dedicated solution. It implies the existence of interfacing features between two parts, for example the keyway in a shaft / key assembly. Also, there must be some sort of hierarchical link between the features of the two parts, so that one acts as reference to the other.

6. APPLICATION EXAMPLE

In the aeronautics industry, products are very complex, often made of some tens of thousands of components. This industry recognizes the capability of CAD tools to improve the quality, lead-time and development costs of its products, as evidenced by its systematic use of digital mock-up technology. The application example discussed hereafter was created as a specific project to assist airframe structure designers at Bombardier Aerospace in the modelling of joggles in order to enhance design productivity.

Figure 3 shows a few typical structural parts that can be found in an aircraft. This specific application of technological links focuses on two target features: joggles and cut-outs. The joggle, commonly found in an aircraft's structure, can be defined as a local depression performed on a part to provide an adequate mating area with juxtaposed parts (*Figure 4*). A cut-out is literally an opening made in a part to accommodate the passing of another part (*Figure 5*, Step 6). With current CAD systems, manual solid modelling of cut-outs and especially of joggle geometry is a long, repetitive and tedious task. Moreover, an aircraft contains thousands of these features. For cut-outs, the manual definition process generally requires profiles to be created by adding some clearance to the geometry of the design context. The profile is also designed to minimize weight and manufacturing constraints. Once designed, the profile is extruded and the resulting solid is subtracted from the main body. The joggle design process also requires the designer to apply knowledge to the design context geometry in order to define construction geometry used to create some solid primitives. These manipulations represent a significant amount of engineering time at the detail design phase, where parts are known to be modified often (the design activity being iterative in its very nature), before being completed and sent to manufacturing. Furthermore, repeated modifications of the models easily lead to product data inconsistencies within the digital mock-up.

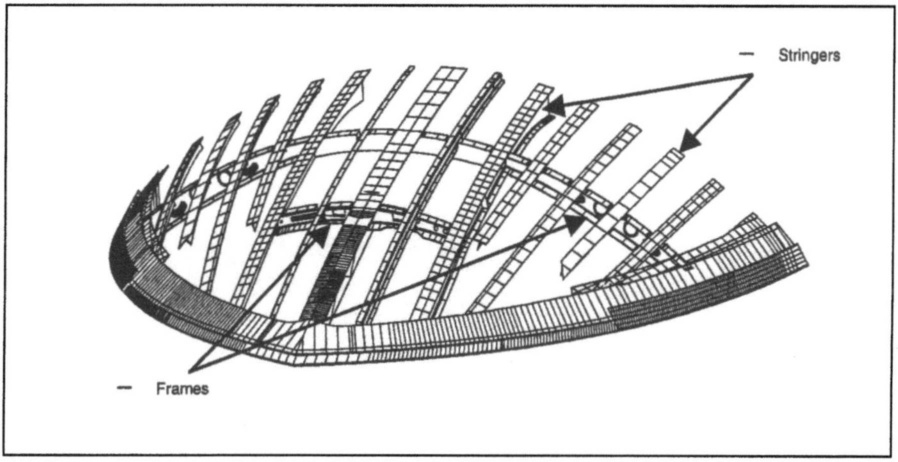

Figure 3. Typical structural parts found in an aircraft

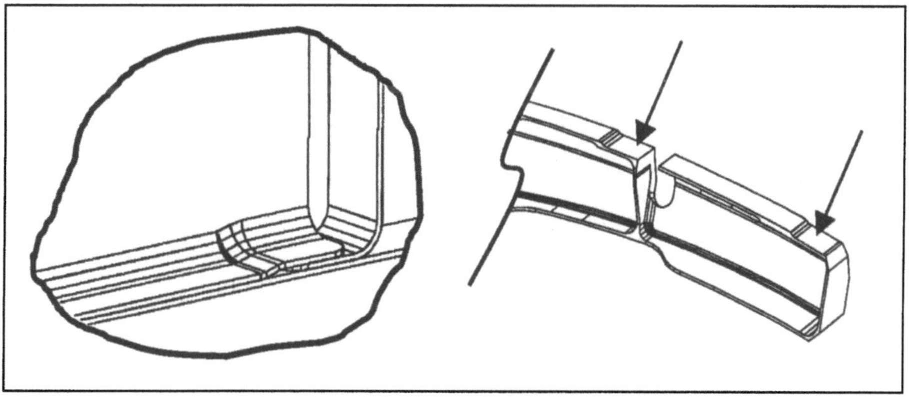

Figure 4. Joggle in a sheet metal part (left) and in a machined part (right)

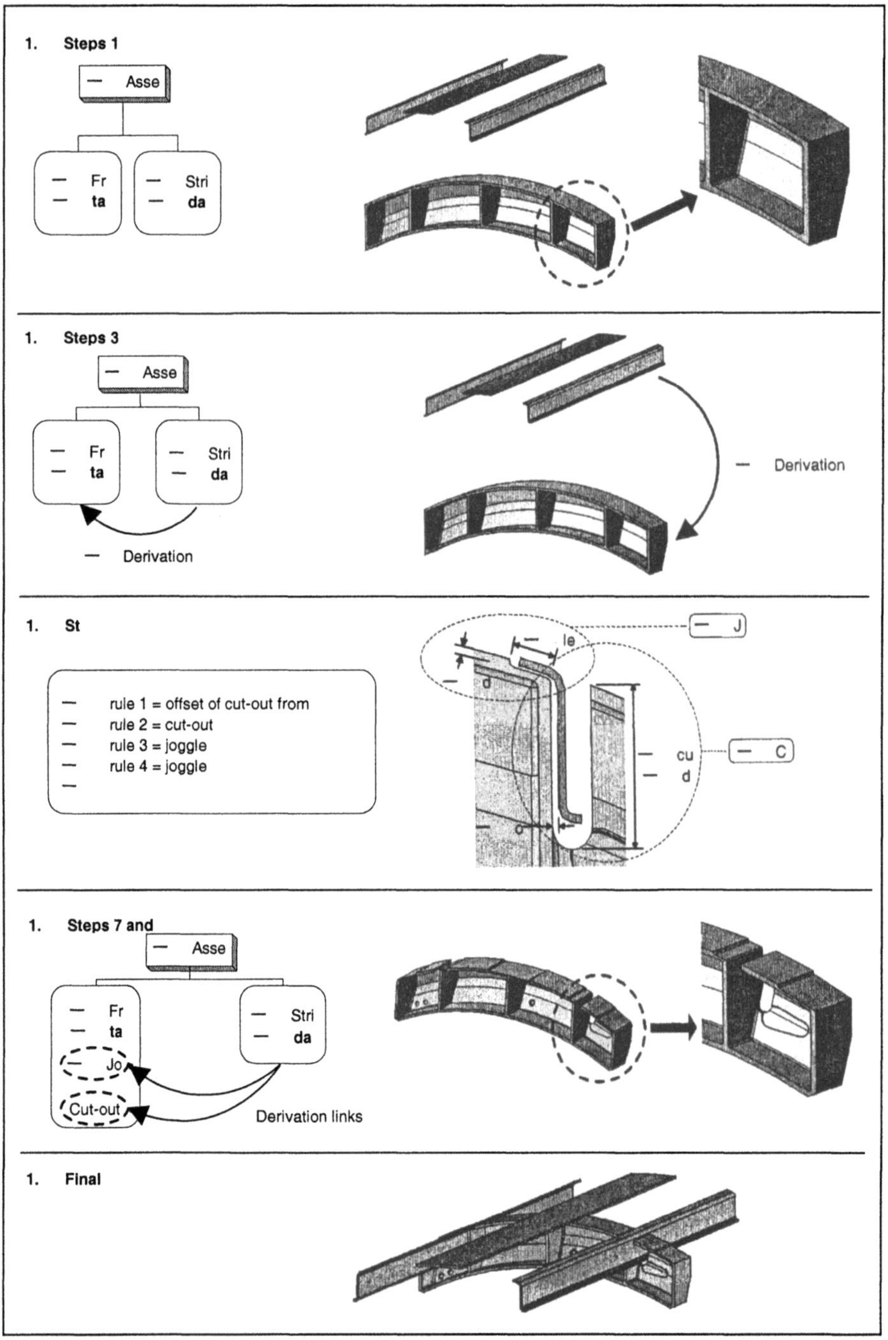

Figure 5 Creation process of a joggle and a cut-out

The great incidence of this problem and since the shape of joggles is mostly governed by the structural parts involved in the design context, makes it a convincing example for the application of the pseudo-imprint and derivation link concepts along with the method presented previously. *Figure 5* illustrates the assisted creation process of a joggle and a cut-out using the proposed method.

In *Figure 5* the first stage represents Steps 1 and 2 of the method, where the designer identifies the joggle/cut-out creation problem, creates the basic geometry of the frame and positions this part in the design context. The second stage represents Steps 3 to 5, during which the derivation link is established once the designer has selected the datum part (the stringer) and the target part (the frame) from the design context. With this technological link, the application determines the relative position and orientation of the stringer in the frame coordinate system. Thereafter, it obtains from the stringer model the independent parameters needed to define the joggle and the cut-out. These values are stored with the derivation link in the application database. Subsequently, in Step 6, the geometry of the pseudo-imprint is chosen according to the analysis of the geometry of the design context and according to the knowledge implemented. The dimensions of this geometry are then calculated by means of specific design rules. During Step 7 the pseudo-imprints are created on the frame, thereby forming the joggle and the cut-out. Finally, Step 8 involves the propagation of all subsequent changes for pseudo-imprint re-evaluation, thus ensuring that the coherence of the assembly is maintained. The latter step is performed when the geometry of the stringer or the position of one of the two parts has changed.

7. CONCLUSION

The proposed framework brings many benefits to some of the modelling tasks found in the design process. Firstly, it provides a net increase in productivity for specific, long, repetitive and tedious modelling tasks. This is achieved in part by allowing designers to interact with the CAD system at a higher semantic level to capture design intent. Secondly, it enables the formalization of design knowledge stored in the application database, such function being triggered by technological links. Lastly, the proposed framework makes for a CAD system that is highly flexible and reactive to modifications in product definition, one of the keys to maintaining product data consistency in the digital mock-up.

In summary, our method exploits the design context by creating a derivation link between two parts in order to obtain specific information from the datum part. Such information is then used to derive the pseudo-imprint by applying specific design rules previously implemented in the application. Subsequently, the pseudo-imprint is created on the target part, and the derivation link is maintained to keep the coherence of the assembly. In so doing, we reach our objective of exploiting the design context to manage product data consistency. The proposed solution is CAD-system independent and uses the top-down approach so as to support the design process from the preliminary to the detailed design phase.

In contrast to generic modelling tools, the proposed dedicated applications make use of specific knowledge to relieve designers from performing tedious tasks, thereby enhancing their productivity. Since dedicated applications are associated to specific design problems, the proposed framework was applied to a specific design task. However, the basic concepts are generic enough to be applied to a collection of other design problems by defining different technological links to capture relationships between two or more parts. More specifically, the proposed approach can be applied whenever repetitive context-dependant geometry need to be defined on parts or features sharing specific common parameters and attributes (therefore constituting a "class" in terms of oriented object programming). As a result, the proposed framework offers strong potential for the development of specific task oriented applications to help designers solve various design problems.

ACKNOWLEDGEMENTS

The work reported in this article was supported by the NSERC (the Natural Sciences and Engineering Research Council of Canada), by the J.A. Bombardier Foundation and by the École de Technologie Supérieure.

REFERENCES

Bordegoni, M., and Cugini, U. (1999). Feature-based assembly in aeronautics design: from concepts development to formalization, *International Journal of Vehicle Design*, V.21, No.2, pp.228-242.

Bronsvoort, F., and Jansen, F.W. (1994). A multi-view feature modeling for design and assembly, *Advances in Feature Base Manufacturing*, J.J. Shah, M. Mantyla and D.S. Nau, Elsevier Science B.M.

Deneux, D. (1999). Introduction to assembly features - an illustrated synthesis methodology, *Journal of Intelligent Manufacturing*, Kluwer Academic Publishers, V.10, No.1, pp.29-39.

Gui, J.-K., and Mäntylä, M. (1993). New concepts for complete product assembly modeling, *Proceedings of the 2nd Symposium on Solid Modeling and Applications*, May 19-21, pp.397-406.

Gui, J.-K., and Mäntylä, M. (1994). Functional understanding of assembly modeling, *Computer Aided Design*, V.26, No.6, pp.435-451.

Libardi, E.C., Dixon, J.R., and Simmons, M.K. (1988). Computer environments for the design of mechanical assemblies: a research review, *Eng. Comput.*, pp.121-135.

Macabies, L., Desrochers, A., Rivest, L., and Maranzana, R. (2000). Liens multi-modèles en CAO - Application au rivetage en aéronautique, *3rd International Conference on Integrated Design and Manufacturing in Mechanical Engineering Forum 2000*, Canadian Society for Mechanical Engineering, May 17-19, Montreal, Canada, Paper No 271.

Mailharro, D. (1998). A classification and constraint-based frame-work for configuration, *Artificial Intelligence for Engineering Design, Analysis and Manufacturing*, V.12, pp.383-397.

Rimscha, M.V. (1990). Feature modeling and assembly modeling - A unified approach, *Proceedings of the IFIP WG 5.2/GI International Symposium*, Nov 8-10, pp.203-213.

Shah, J.J., and Mäntylä, M. (1995). *Parametric and Feature-Based CAD/CAM*, John Wiley & sons, pp.173-186.

Shah, J.J., and Tadepalli, R. (1992). Feature based assembly modeling, *Proceedings of the 1992 ASME International Computers in Engineering Conference and Exposition*, Aug 2-6, pp.253-260.

Shah, J.J., and Rogers, M.T. (1993). Assembly modeling as an extension of Feature-Based Design, *Research in Engineering Design*, V.5, pp.218-237.

Sodhi, R. (1991). Representing tolerance and assembly information in a feature based design environment, *Advances in design Automation*, V.1, pp.101-108.

Ullman, D.G. (1997). *The mechanical design process*, 2nd edition, The McGraw-Hill Companies, Inc.

Feature-based Integration of Product, Process and Resources

S. HAASIS[1], D. FRANK[1], B. ROMMEL[1] and M. WEYRICH[2]
[1] *DaimlerChrysler AG - Ulm, R&T, Germany*
[2] *DaimlerChrysler AG - Stuttgart, IT, Germany*

Abstract: The processes currently deployed within the organization of today are being greatly impacted by the demand for drastic reductions in lead-times accompanied by cuts in development and manufacturing costs. Furthermore, the trend towards geographic separation of development and manufacturing facilities and the growing requirement that external partners be seamlessly integrated into internal processes, together with increasing globalisation in manufacturing which necessitates that product design take - often very different - manufacturing conditions into consideration, make the efficient design of both internal and cross-organizational process chains, i.e. comprehensive, complete digital product and production planning, a necessity. The use of features as carriers of descriptive and semantic information within the product development process has led to a substantial decrease in redundant data input and, thus, to a significant reduction in effort and the probability of errors occurring while, at the same time, supporting a product design which is aligned with the manufacturing processes deployed. Supplemented by process data and methods geared for the specification of production aspects and workflows, feature-based process models enhance the transparency of manufacturing capability and resource modelling, thus serving as the foundation for the introduction of new organizational concepts such as simultaneous or concurrent engineering. This work describes the role features play in an organization's overall process chain: the connection between design, manufacturing and measuring features and their consolidation to a backbone for the continuous flow of information for the complete modelling of the product, the process and the resources.

Key words: Feature-based process chain, design features, manufacturing features, resource planning, cost estimation

1. INTRODUCTION

In the past, we typically considered isolated parts of the overall process chain because we were basically concerned with implementing improvements within a certain section or extending single parts of the chain. Yet, the current trend towards taking a comprehensive look at the overall process chain means that, instead of solely ensuring continuity of data to effect process improvement, we now require a uniform, digital representation of all of the systems comprising the process in a single, common data and process model. Whereas, exclusively the relationship between the product specification and the process description used to be relevant for process improvement, step by step, the circle has now come fully around to include the description and integration of the resources utilized, as illustrated in *Figure 1.*

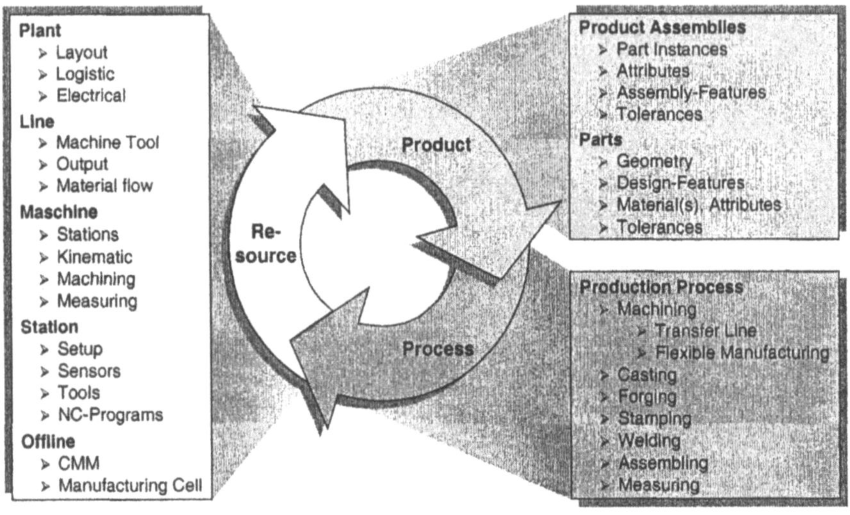

Figure 1. Connection between the product, the process and the resources (Olling 1999)

The use of features as a uniform information medium throughout the process chain is the foundation upon which we base any description of what takes place at any step in the overall process. And the goal of our approach is the complete digitisation of all of the areas concerned in the product creation process (Haasis 1997).

Another reason for taking one step back and looking at the product, the process and the resources as a whole is the corresponding cost estimation. The benefit is that we become capable of making initial cost forecasts at extremely early stages in the process chain by supplementing the entire data network with standard parameters during individual design and NC planning phases, thus enabling early cost optimisation of the individual process phases and ensuring that cost targets are met.

2. FEATURE-BASED PRODUCT DEVELOPMENT

The process chain starts with product development. Here, in the first link, the first features are insinuated into the data flow. The availability of design features in the product development phase is the basis for the continuous and comprehensive improvement of the overall product creation process.

What is basically a rudimentary description of the part to be developed is split up into semantically specified partial models, where the geometry of the part is described on the basis of design features, and several design engineers working in concert create the part, e.g. a cylinder head. In a main document, the various partial models are linked together to form an overall model of the complete part. The overall model may be updated after various modifications have been made to the partial models. One of the partial models contains all of the machinable design features. The other partial models combine to describe the form of the unmachined part (Haasis 1999).

Design Features. This group of features contains general feature types such as a blind thread or a through bore, on the one hand, and product-specific feature types like spark plug boreholes for gasoline engines or glow plug boreholes for diesel engines on the other. And design engineers are supported through a kind of intuitive product modelling geared for their specific work-related activities (Haasis 1997). In particular, the functional perspective as perceived by the designer is supported. Each design feature may be organized according to its function and, as such, is offered to the designer for integration into the product. This way of thinking where the feature is linked to its function promotes the creativity of the designer. General design features may be accessed by all designers working within an organization and are standardized throughout the concern (Haasis 1996, Haasis 1995).

The types of features cited above are supplemented by features belonging to assembly work, which are not restricted to a specific part. For example, the assembly feature "screwed part" assigns the design features "through bore" to the plates to be joined through the screwed connection and the standard parts required for this work such as the hexagonal screw, the locking ring and the hexagonal nut. In addition, these features support subsequent assembly planning and logistics operations.

A feature editor was developed for CATIA V5: the software module integrated in the CAD system supports the designer in functional and intuitive product modelling by specifying and storing the feature on the basis of the function to be mapped. *Figure 2* shows how the dimensioned blind thread is mapped to the function "positioning".

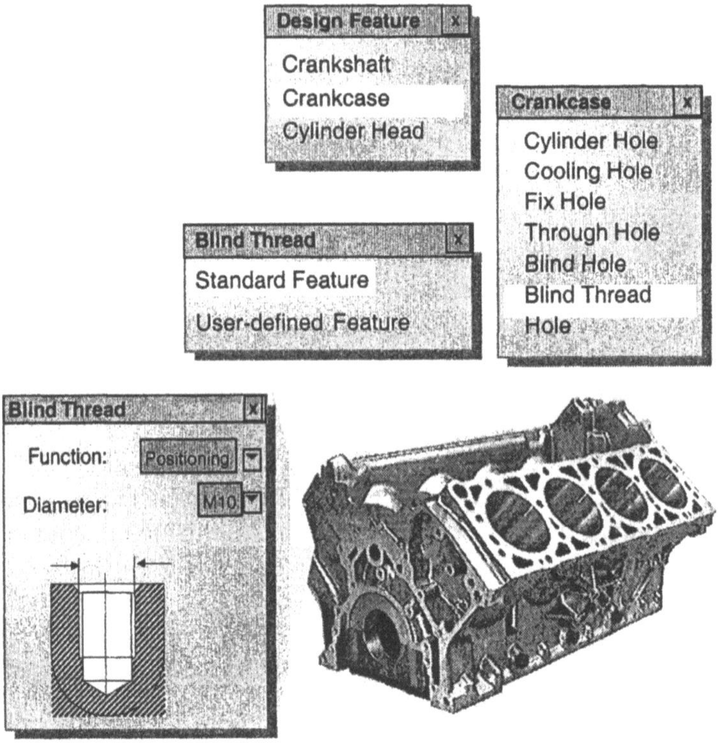

Figure 2. Feature-based product design

The provision of such feature libraries and their deployment considerably accelerate the overall design process. Also, use of standard features ensures efficient product modelling and contributes to enhanced product quality. Designers are provided with tried and tested combinations of parameters which are aligned with the corresponding manufacturing context. Moreover, design engineers are provided with additional user-defined, non-standardized features which they may utilize as needed for their specific design task (Haasis 1997).

While fine-tuning the product model, designers have access to a feature-based cost information system. The result of the analytical approach on the base of a generative process plan is a detailed and differentiated cost estimation that enables us to draw specific conclusions about the cost driver and the possibilities to adjust product cost. Using the product model the generative process plan is derived. It names the machining operations and for each operation the necessary resources and the machining time. The cost model evaluates monetarily the machining time using the corresponding hourly rates. Generating the process plan in (Kiritsis 2000) and (Ou-Yang 1997) a rule-based system is used to assign machining operations and resources to features. However, manufacturing technology and the resources available underlie a continuous improvement and change and, machining strategy is often company specific. The methodology of case-based reasoning (CBR) makes it possible to adapt to new manufacturing technology and to consider experience knowledge. (Paulokat 1994) describe the planning for machining workpieces using CBR and (Rehman 1998) use CBR in the field of cost estimation.

The example shown in *Figure 3* illustrates how the designer is assisted in evaluating the costs which the digital product model would incur, effecting early cost calculations and meeting targeted costs. Hence, features are vital for ensuring a design process geared for manufacturing and cost efficiency.

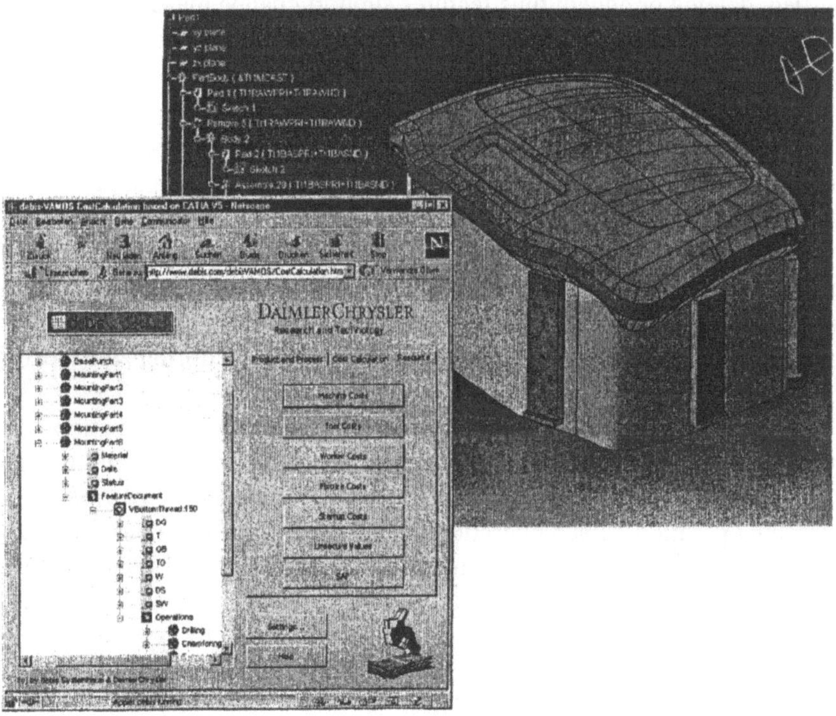

Figure 3. Feature-based cost estimation

Now that the product model has been completed, a model of the related manufacturing process is required. Depending on the maturity of the product model and supported by a workflow management system, the work scheduler generates a process plan based on the features used in the product design (Weyrich 1999). The exact role played by features within this sub-process is set out in the following section.

3. FEATURE-BASED PROCESS DESIGN

This part of the process chain comprises the modelling of the product manufacturing and inspection on the basis of the features used during product design. We will refer to the applications following the design phase as downstream applications. The CAD model consisting of design feature instances is transformed into several downstream-application-specific feature models. If a design change should be made, the downstream applications are given exact information on any features which have been modified, deleted or added; thus features support change management.

3.1 The manufacturing perspective

To create manufacturing models automatically from the design models a process called feature mapping is used. *Figure 4* shows as an example design models for the part and its casting tool.

The design models are mapped onto feature-based manufacturing models and inspection models which constitute the specific experts' view on the product or part respectively. These specialists will create manually those elements which cannot be generated automatically during the feature mapping.

The next planning step adds technology-specific information. For instance, when planning the machining process in more detail, the NC planner assigns to the machining features a sequence of individual machining operations. Each operation is, in turn, assigned to exactly one tool that will be used for its execution (Storr 1996, Geissler 1996, Schuetzer 1995).

The system will suggest the operations and tools to be used for the individual machining features. The NC planner reviews these solutions and, if need be, may alter them.

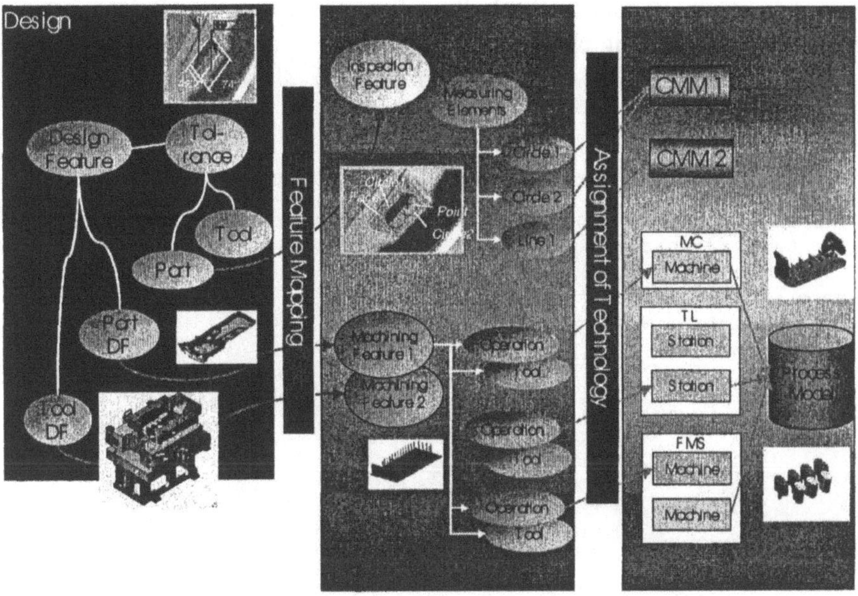

Figure 4. From design to manufacturing (machining, casting) and inspection

3.2 The process perspective

The workpiece- or manufacturing feature-oriented structure of the machining model does not contain a suitable sequence of operations for the subsequent manufacturing process. So, a process perspective of the overall machining model has to be generated in parallel to the manufacturing perspective. And, in contrast to the manufacturing perspective, the process perspective consists of a cross-feature sequence of individual operations, each of which is allocated a specific workstation within the manufacturing process.

The scope of operations is distributed to fixtures where several operations according to the requirements which the machining of the workpiece places on the manufacturing process are carried out. Each of these fixtures may incorporate more than one NC program. The operations are divided up between the NC programs to ensure that the machining technologies run in the appropriate order. Before an NC machining program can be generated for such an NC program, a suitable NC machine or workstation has to be specified. Thus, an unbroken link between the design features, the manufacturing features and, finally, the machine to be used for the operations performed in connection with a machining feature is formed. This continuous chain of data is the bottom line when determining machining costs during the planning phase. The approaches mentioned above have been implemented in CATIA version 5. *Figure 5* shows the CATIA V5 Manufacturing GUI.

Figure 5. Feature-based process planning

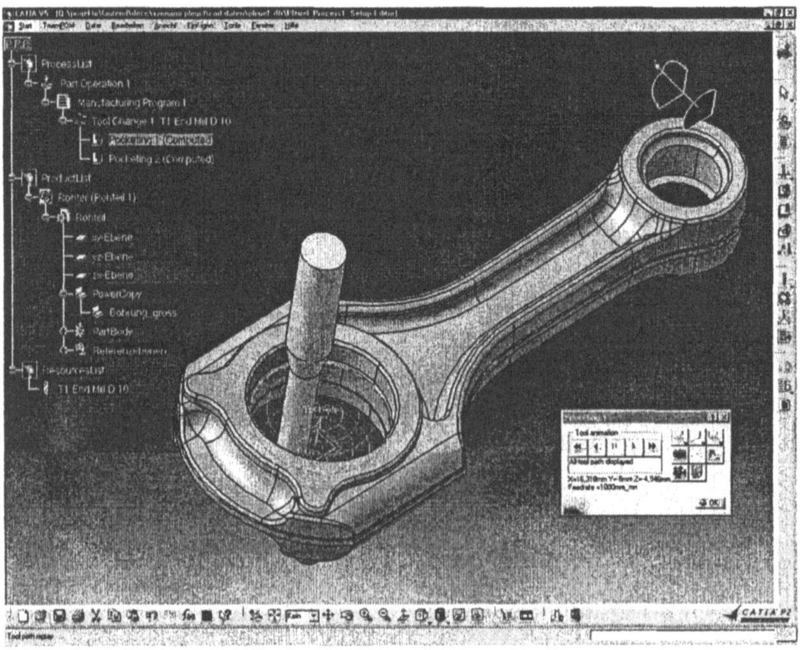

3.3 The same procedure for measuring operations

Similar to the procedure described for the machining model is that used for measuring operations. However, naturally, measuring features are generated for the relevant shape elements. And the measuring features are assigned one measuring method and one measuring sensor each during planning of measuring operations. Just as in machining, the workpiece-oriented perspective is separated from the process perspective. Where machining techniques and tools are assigned to the machine centres and workstations making up a transfer line, the measuring methods and sensors are assigned to the coordinate measuring devices and measuring stations selected in the line. When the product and process descriptions have been completed, resource planning is effected on the basis of the equipment and facilities required for the manufacturing process.

4. FEATURE-BASED RESOURCE PLANNING

When planning the deployment of manufacturing equipment, engineers use both the existing design results in the form of 3-D models and 2-D drawings and context-relevant product requirements to first create the product specifications. As an integral part of the design results, features, in addition to providing the parts making up the form of the component, yield information on casting aspects, tolerances and other technological requirements. Taking these requirements into account, the best suited machining operations and tools are selected and the resulting machining volume decided. This, in turn, impacts the layout of the stations, machines and lines. Cycle and machining times are considered and the material flow and line layout planned, *Figure 6* Here, close cooperation with suppliers and system partners is required, necessitating their integration into this process.

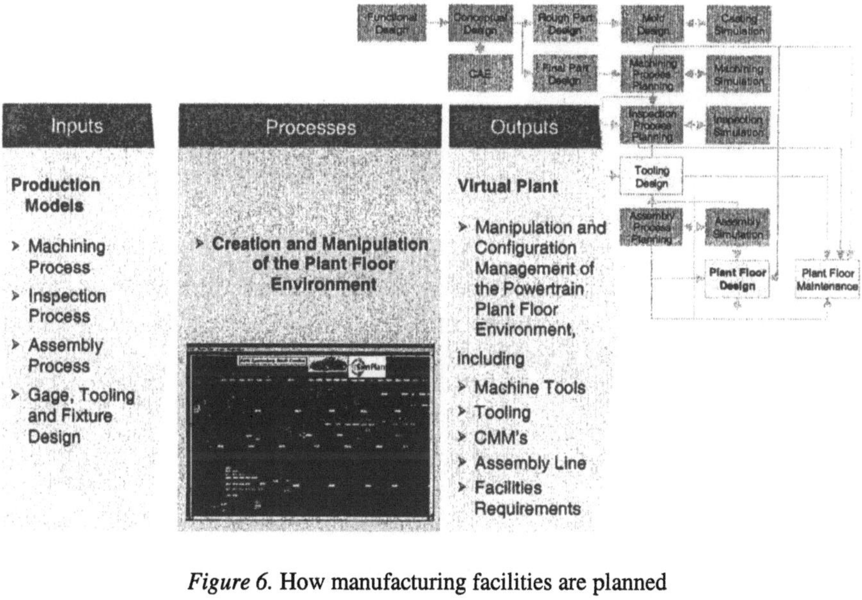

Figure 6. How manufacturing facilities are planned

As soon as the sections of the manufacturing line have been planned by assigning the operations and tools needed and the resulting machining volumes to the stations and machines in the line, a feature-based manufacturing process model may be generated for each resource with the objective of providing a representation of the respective status of the workpiece being machined. This is generated, if required, through subtraction of the 3-D geometry model - unmachined part or status after machining at the previous station - from the sum of the remaining machining volumes at the station. In addition to enabling navigation of the digital manufacturing of the part based on the visualized representations of the machining status before and after it has run through a station, these models permit situation-related offline programming of coordinate measuring devices (Haasis 1999). See *Figure 7.*

Figure 7. Planning manufacturing stages by using 3-D manufacturing process models

Suitable techniques still remain to be developed to facilitate integration

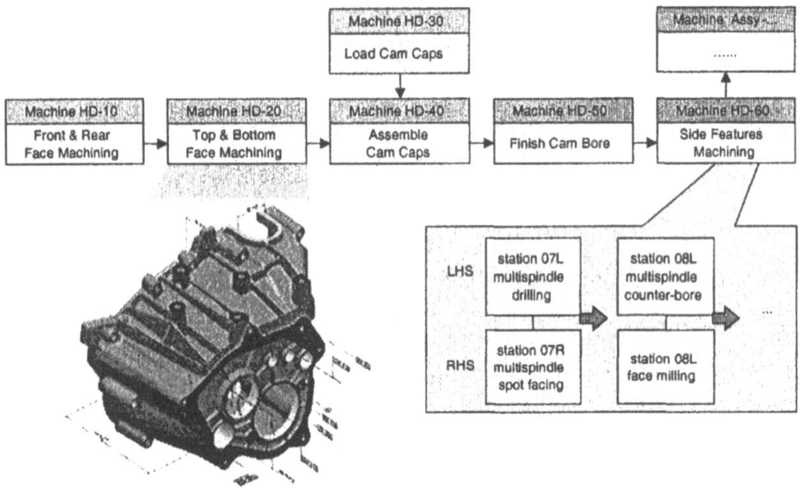

of the approaches described in existing process chains. The chapter set out below shows how the approaches might be implemented in real process chains and describes current project work in this field.

5. IMPLEMENTATION IN REAL PROCESS CHAINS

5.1 Objectives for the realization of the data model

The integration of process chains is an essential task if changes to processes are to be made quickly and with optimal resources. In order that single links in the process chain may be more easily integrated, a new generation of IT tools is required to enable effective data exchange. In this connection, a performing change management throughout those systems of the process chain concerned is a necessity for the practical implementation. Hence, a uniform data model to be deployed throughout the process chain for quick and easy data exchange between the individual processes has become a top priority.

Features offer a performing methodology for the encapsulation of information and its exchange as a logical unit at the process seams. And continuity of the process chain in the case described in the previous chapters is given through the linking of data from design, process planning, prototype construction and mass production. Of course, features need to be structured appropriately to ensure that they are suitable carriers of the information generated in the different processes. Hence, the challenge we face is how to design and implement structured features as building blocks for the transfer of data throughout the process chain to enhance continuity and control in the long run (Haasis 1996).

5.2 Process chains

Today's process chains are characterized by a strict demarcation between sub-processes, which represents a barrier to the flow of information. The borders between the individual processes are historical in origin and their continued growth is typically a result of the organizational structure of the concern (Haasis 1996).

In the conventional process chain as frequently implemented throughout industry today, there are two barriers to the data flow, i.e. there are three separate and distinct process areas. As a rule, the first area is design, where CAD tools play the major role in the generation of data. Design is clearly separated from the second area, prototyping, and its related data generation from NC programming. In prototyping, small lots are produced at machining centres programmed using special CAM tools. After NC machining and prototyping have been successfully completed, mass production of the product - the third area mentioned - can begin. In mass production, transfer lines are set up for the manufacturing of very large lots in contrast to the machining centres used in small-lot prototype manufacture. This represents the second block to the data flow, as mass production is controlled by entirely different machine programs customized for transfer lines. Unfortunately, in practice, design changes often need to be made after commencement of mass production, thus necessitating corresponding changes in all the areas concerned from the beginning of the process chain to the end at the machine. The result is a drastic increase in the time and effort required.

In flexible manufacturing - supported through feature technology - the barriers between the process parts are torn down. However, this can only be done if the digital transfer of data is warranted from the beginning of the process chain - design - through prototyping and on to the manufacturing process itself. In principle, standard formats (e.g. STEP) would seem a suitable way of effecting the exchange of data. However, practice has shown that single-vendor CAx system solutions where system modules are implemented in the process areas involved are the better alternative since they enable efficient data exchange within a single system environment.

Implemented on top of a CAD/CAM platform in place within a process chain, feature technology would form an excellent basis for the compilation of information and its transfer in the form of a logical unit over the border from design to NC programming. Although the form in which the features are represented would, of course, have to change due to the very different perspectives inherent in design and manufacturing, a great deal of data content could be transferred as is, without change.

The second barrier to data flow is the exchange of NC programs between the machines used in prototyping and those deployed in mass production. With ISO 14649, new machine programming standards are in the pipeline for the improved data exchange between machines of different types. Here also, feature technology would provide a suitable means of facilitating data transfer by enabling the creation of NC programs which are independent of the type of machine and the control system deployed.

5.3 Implementation

The process chain as set out in the previous sections will be migrated to a chain supported by features. To solve the underlying problem of data continuity, DaimlerChrysler has entered into a strategic partnership with the French-based systems house Dassault, establishing the CAx system CATIA as the system of choice organization wide. In spite of the restriction this puts on the possible options open for system deployment, continuous data exchange from the beginning to the end of the process chain is ensured due to this single-system solution. Also, this framework has enabled activities pertaining to continuous feature processing to be carried out, with the concrete goal of establishing a feature library tailor-made for the CAD and CAM applications deployed within the organization.

The Process Development Department in Untertuerkheim bore the overall responsibility for integrating the system at DaimlerChrysler. The system is based on CATIA V5 and is made up of three major components (Haasis 1997):

- A feature editor as a CAD module for workpiece modelling, including a feature library.
- A CAM module for intuitive NC programming.
- Extended functionalities for flexible manufacturing and simulation.

The components are currently implemented as prototypes for drilling and milling applications and are deployed in product industrialization. Further modules for NC machining simulation and workpiece fixture planning are in the pipeline.

The prototype as implemented in the process chain has served to evidence the great potential inherent in feature technology: features have proved their suitability for uninterrupted data flow throughout the process chain. At the same time, we have found that the related feature libraries set up need to be carefully structured and tailor-made to the processes in place within the organization or they will not be widely accepted by the users.

6. CONCLUSION

Feature technology comprises the deployment of design, manufacturing and measuring features in their respective areas and serves as a backbone for the bi-directional flow of information throughout the overall process chain, thus enabling continuous, digital product and production modelling (Haasis 1997, Haasis 1999). Should any changes be made in design in a feature-based process chain, the resulting impact on the costs incurred, the machining technologies to be adapted, including utilization capacities, and the quality of the product are quickly and easily derived. The provision of standard features in application-specific feature libraries allows standardization and ensures the desired quality of the product description. In addition, direct access to these standard libraries accelerates the design process and permits initial cost estimations as early in the overall process as the design stage. When standard features are brought into alignment with manufacturing, the stocks of tools required for production can be right sized, leading to further cost cutting. In NC programming, the machining techniques used are closely coupled to the manufacturing features, thus enabling the creation of a feature-based object-oriented NC program, which, through the development of a new generation of NC controls, may be directly read and edited in the future (Haasis 1993). The distribution of machining techniques to the NC machines as needed will enhance the utilization capacity. And, similar to the creation of NC programs where machining techniques and tools are assigned to manufacturing features, feature technology supports measuring programming (Haasis 1994) by assigning measuring techniques and tools to measuring features. See *Figure 8*.

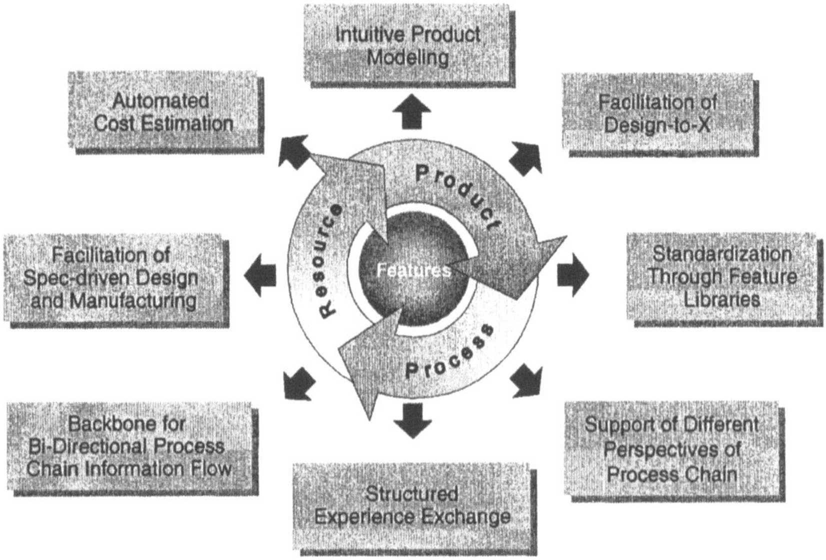

Figure 8. Benefits expected through the deployment of feature technology

Initial integration tests have evidenced the potential inherent in feature technology (Haasis 1997):

- Feature technology accelerates generation of product and manufacturing descriptions.
- Feature technology leads to the efficient creation of NC programs, their optimisation and to efficient change management.
- Feature technology enables integration of CAD/CAM data by means of integrated data administration in CATIA V5.

Yet, the full potential of feature technology can only be exploited if feature-based workflows are implemented throughout the process chain. Partial utilization of features in parts of the process chain results solely in partial benefit. Only through the continued, consistent transfer of feature technology to the workflows in the process chains deployed within the DaimlerChrysler AG business units will the maximum yield be reaped.

REFERENCES

Gao, J. X., Huang, X. X. (1996). Product and Manufacturing Capability Modelling in an Integrated CAD/Process Planning Environment, *Int. Journal of Advanced Manufacturing Technology 11,* No. 1, pp. 43-51.

Geissler, K., Wilfert, H. G. (1996). Generative Arbeitsplanung in der Fertigungsindustrie, *ZwF 91*, No. 3, pp. 105-108.

Haasis, S. (1997). Feature-basierte Prozesskette mit Object-NC. In *Technologietage 1997 der Daimler-Benz AG.*

Haasis, S. (1998). Feature-Based Process Chain within the Scope of Powertrain Engineering. Proceedings of the 31st ISATA, 2nd – 5th June 1998, Duesseldorf, pp 113-120.

Haasis, S. (1997). Handlungsorientierte Gestaltung von CAx-Systemen entlang der Prozesskette. In *KIK '97*. Konstanz: Universitaetsverlag,.

Haasis, S. (1997). Nutzenpotentiale der durchgaengigen Feature-Verarbeitung, *VDI-Berichte Nr. 1322*, VDI-Verlag, Duesseldorf, pp. 63-82.

Haasis, S., Bär, T.; Pfeifle, J.; Rothenburger, G.; Schulze, H.; Ströhle, H.: *User-Oriented Inspection Model for Powertrain Engineering. Science Days 2000, Sindelfingen.*

Haasis, S., Katzenbach, A.: *Engineering Processes of the Future. Digital Products - Living Data is the Future. Page 65-73. Teubner, Stuttgart: 2000.*

Haasis, S., Rose, H., Schulze, H. (1999). *Technische Unterstuetzung erfahrungsgeleiteter Arbeit mit Prozessketten in der Produktion,* In Innovation und Kooperation – Nutzerorientiertes Konzept fuer Interaktionssysteme in der Serienfertigung., Campus Verlag, Frankfurt u. New York.

Haasis, S.; Schulze, H. (1997). Erfahrungsgeleitete Arbeit mit Prozessketten in der Produktion, *1. Kassler Kolloquium.*

Kiritsis, D., Xirouchakis, P. (2000). DEVIPLAN: A bid preparation system for mechanical SME, In *Proceedings of DETC2000.*

Meier, M., Bichsel, M., Elspass, W., Leonhadt, U., Wohlgesinger, M. & Zwicker, E. (1999). Neuartige Tools zur effizienten Nutzung der Produktdaten im gesamten Produktlebenszyklus. *Konstruktion* (9), 11-18.

Olling, G.; Nedwick, J; Scharpf, J. (1999). Common View of CAD/CAM Technology Integration in Powertrain Processes. DaimlerChrysler: Stuttgart and Auburn Hills.

Ou-Yang C., Lin, T.S. (1997) Developing an integrated framework for feature-based early manufacturing cost estimation, *Advanced Manufacturing Technology,* pp. 618-627.

Paulokat, J., Wess, S. (1994). Planning for machining workpieces with a partial-order, nonlinear planner, In: *Proceedings of: AAAI Fall Symposium "Planning and Learning: On to Real Applications".*

Rehman, S., Guenov, M.D. (1998). A methodology for modelling manufacturing costs, *Computers and Industrial Engineering*, Vol. 35, No. 3-4, pp. 623-626.

Salomons, O. W. D., Geelink, R., Van Slooten, F., Van Houten, F. J. A. M. ,. P. & Kals, H. J. J. ,. P. (1998). Definition of design and manufacturing form features *[Twente-Literatur]. In Proceedings of 31st CIRP International Seminar on Manufacturing Systems; Networked Manufacturing: Integrated Design, Prototyping and Rapid Fabrication, May 26-28, 1998, Berkley, California .*

Shah, J. J. & Mäntylä, M. (1995). Parametric and feature-based CAD/CAM. Concepts, Techniques, and Applications. New York: John Wiley & Sons.

Shah, J. J. & Yan, Y. (1996). Representation and Mapping of Geometric Dimensions from Design to Manufacturing. In *Proceedings of the 1996 ASME Design Engineering Technical Conferences and Computers in Engineering Conference, August 18-22, 1996, Irvine, California* (S. 1-11).

Shah, J. J., Bhatnagar, A. & Hsiao, D. (1988). Feature Mapping and Application Shell. In *Proceedings of ASME Computers in Engineering Conference* (S. 489-498).

Shah, J. J., Yan, Y. & Zhang, B-C. (1998). Dimension and tolerance modeling and transformations in feature based design and manufacturing. *Journal of Intelligent Manufacturing, 9* (5), 475-488.

Storr, A., Rommel, B. (1996). Bearbeitungsobjekte als Basis fuer Informationsmodelle. *In Objektorientierte Produktionsarbeit: neue Konzepte fuer die Fertigung.* Campus Verlag, Frankfurt u. New York.

Van Diepen, A. (1993). *CAD-CAPP integration for sheet metal products [Master of Science Thesis].* Twente (NL): University Twente.

Weyrich, M. (1999). *Multimediale Werkzeuge zur 3D-Planung im Maschinen- und Anlagenbau.* VDI-Verlag, Duesseldorf.

Weyrich, M.; Rommel, B.; Haasis, S.; Müller, P.: *First Prototype of a NC Controller Based on STEP-NC. Science Days 2000, Sindelfingen.*

Wolfram, M.(1994). *Feature-basiertes Konstruieren und Kalkulieren,* Hanser , Muenchen u. Wien.

Product Model Based Multiple Viewpoint Information Sharing Using Features Technology

J.O. CANCIGLIERI[1] and R.I.M. YOUNG[2]
[1]*Pontifícia Universidade Católica do Paraná – LAS Laboratory of Automation and Systems – Rua Imaculada Conceição, 1155 – Prado Velho – Curitiba – PR- Brazil – CEP 80215-030 – e-mail: osiris@rla01.pucpr.br.*
[2] *Wolfson School of Mechanical and Manufacturing Engineering - Loughborough University - Loughborough, Leicestershire - LE11 3TU UK – e-mail: R.I.Young@lboro.ac.uk.*

Abstract: Concurrent engineering is an important philosophy in achieving better time to market in new product development. While the use of design teams is achieving some success there is a need for modern software tools which support the design process to be radically improved. This paper provides a contribution in the area of decision support systems based on the use of product and manufacturing models to provide appropriate information. Typically design for manufacture systems provide only support for one process e.g. design for assembly, design for machining. However, when the full breadth of design for manufacture is considered there are many aspects which must be addressed. This leads to the need for information systems to be able to support multiple views of a product, where each view provides the appropriate representation to support one manufacturing perspective. This paper discusses the requirements of a reasoning system which is able to provide multiple views of a product and which can provide mechanisms to move from one viewpoint to another. The research focuses on injection moulding and considers viewpoints such as product functionality, product mouldability, mould design elements and mould manufacturing viewpoints. A case study product is discussed to illustrate the approach being used.

Key words: Concurrent Engineering, Features Technology, Machining, Product Modelling and Product Mouldability, Design For Manufacture.

1. INTRODUCTION

A crucial problem in the development of new products is to reduce the time required for their design and manufacture while still maintaining high quality and minimum development cost. To achieve this inputs are needed from experts in a multitude of disciplines as well as from customers and suppliers. The overall design process must be well orchestrated and integrated. Hence the interest in recent years in Concurrent Engineering.

Concurrent engineering is an important philosophy in achieving better time to market in new product development. However, while the use of design teams is achieving some success, there is a need for the software tools which support the design process to be radically improved. One important aspect of Concurrent Engineering is Design For Manufacture (DFM). This recognises that a company cannot meet quality and cost objectives with isolated design and manufacturing engineering operations. Thus, the DFM approach embodies underlying manufacturing imperatives that help to maintain communication between all components of the manufacturing system and enables the design to be adapted during each stage of product realisation (Boothroyd et al. 1993). Other imperatives include a general attitude that resists making irreversible design decisions before they absolutely must be made and a commitment to the continuous optimisation of products and their manufacturing processes. Therefore, the objectives of the design for manufacture approach are to identify product concepts that are inherently easy to manufacture, to focus on component design for ease of manufacture and assembly, and to integrate manufacturing process design and product design to ensure the best matching of needs and requirements.

Design systems should integrate Design, Manufacturing, database and expert systems tools (Willems *et.al.* 1995). Such systems should aim to assist concurrent engineering teams by providing access to consistent sources of product and manufacturing information, and providing applications that support decision making.

Feature technology has been expected to provide a better approach to the integration of design and the manufacturing activities following design such as engineering analysis, process planning, machining, fixturing, etc. (Salomons *et.al.* 1993). However, researchers have explored the use of features technology with only limited success. Typically, features have been used to represent one design or manufacturing viewpoint e.g. form, machining, assembly, etc. To be successful, integrated software systems must link a range of different views of products e.g. geometric view, functionality view, mouldability view, manufacturing view and so on. Current approaches are incapable of meeting these needs (Allada, and Anand 1995).

This paper explores the potential of utilising features based approaches in combination with information modelling techniques to provide comprehensive sets of information which can support a range of manufacturing viewpoints. Each information set can be seen as supporting one aspect of decision making and will refer to aspects of form or other attributes of a part. These sets can be used in reasoning about the design, or performance in manufacturing the part.

This paper provides a contribution in the area of decision support systems based on the use of product and manufacturing models to provide appropriate information. Also, It proposes a reasoning system structure which can cope with the interactions between different views of the product i.e. product manufacture, product design, product mouldability, mould design and mould manufacture.

2. MULTIPLE VIEW DESIGN AND MANUFACTURING

Design for Manufacture software systems of the future must provide support for multiple processes, because design for manufacture involves considerations of many different perspectives of the product, so it is increasingly necessary to balance and integrate work with multiple views (or perspectives) of the product. This view is supported by the work of (Rosenman and Gero 1996) who have explored a multiple view approach in architectural design.

One of the approaches to multiple view design for manufacturing is based on the use of agent systems. Agent systems have been defined differently by researchers, but definitions which are suitable in this context are "the agent system is viewed as a system capable of general intelligent action" or it is a "type of knowledge-base system" (Wooldridge and Jennings 1995). However, the communications or translation of information between distinct domains still causes a problem due to the diversity of types of information involved in a multiple view environment.

The development of design for manufacture applications has, typically, been focused only on single manufacturing viewpoints (Gupta et. al.1997, Krause et al 1995, Boothroyd 1994 and Fauvel 1994). For example, (Gupta et al. 1997 and Sanchez et al. 1997) approaches design for manufacture solely from the machining perspective and (Boothroyd 1994) concentrates on design for assembly. From these types of work it can be seen that each view taken requires its own structure of product information. Furthermore, the structuring of the product information is similar to the structuring pursued in the area of features technology, e.g. machining, assembly, fixturing features and so on.

Product Modelling is another important aspect of research involving the development of CAE systems. It has been recognised by both research and application communities over the past several years that an advanced information system is needed to integrate and co-ordinate the various life-cycle considerations during product realisation (Krause et al. 1993). For more than a decade, the International Organization for Standardization (ISO) has been working on developing the STEP standard which is strongly focused on data exchange (between systems).

It is also important for Product Models in the future to be able to deal with multiple view representations of the product. An important aspect of research into information driven CAE systems is therefore to develop product models, which can capture different perspectives of the product. This enhanced system functionality can be achieved through the sharing of information related to representations of different aspects of the product in the product model (Fowler 1995). Recently, the STEP community has also recognised that there is the need to share information and integrate different viewpoints of products as evidenced by the work of group WG10 (*ISO TC184/SC4/WG10 N219*).

Product model research is typically concerned with the definition of the structure of product information, which drives a particular application. The MOSES research, developed between Loughborough and Leeds Universities (UK), has dealt with multiple applications, supporting design for manufacture across machining and injection moulding and also covering applications in design for function and manufacturing planning. However, each of these application areas has been supported by an independent single view of the product. In a multiple viewpoint design and manufacture system such views of the product must be considered to be interdependent. *Figure 1* provides a representation of MOSES concept with the overlay, represented by the question marks, of the need to interrelate the information, which supports each application.

This has pursued issues related with multiple viewpoints in a design and manufacturing environment by addressing two related questions:

1. What structure of information in the product model is needed to support different but related views of the product?
2. How can the relationships between these sets of information be defined?

In this work the authors have considered only the relationships between sets of information and they have been targeted at the area of injection moulding to provide an appropriate scope for the research. The next section will present the issues related to the area chosen.

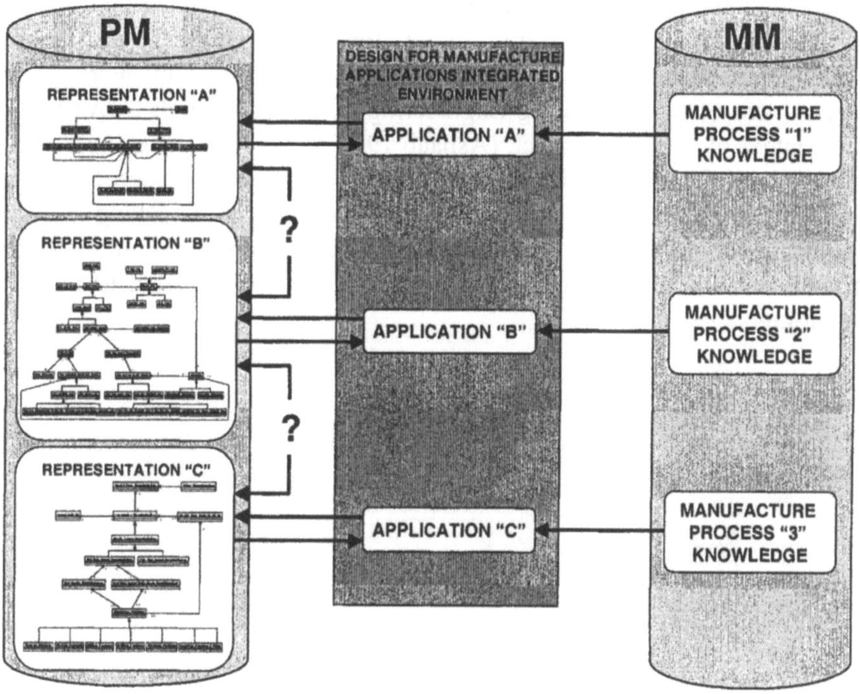

Figure 1. CAE system architecture to support multiple viewpoints.

2.1 General issues in multiple viewpoints in injection moulding

Injection moulding is a huge area that covers several expertise sub-areas. This area has opened opportunities for researchers to explore the problems in multiple-view points using basically features technology due to the diverseness and complexity of the issues related with plastic part, mould and manufacturing of them. (Pratt et al. 1993) have developed a knowledge for assessment of the manufacturability and processability of the part. But their research has been constrained only to design the plastic part. (Weinten and Manoochehri 1996) have presented a methodology only for defining the parting line location based on the geometry of the plastic part.

(Chin and Wong 1996) have developed a Knowledge-base system (KSB) for designing plastic products placing emphasis on the conceptual design stage. (Lee et al. 1997) have presented a systematic methodology and knowledge base for injection moulding mould design in a concurrent engineering environment. The approach taken is concerned only with the design of the mould. However it does not show how the geometry of the plastic part can be transferred to the mould design.

In the area of CAE systems support for the design of plastic injection moulds, (Ashaab 1994 and Lee 1996) have carried out work at the Department of Manufacturing Engineering of Loughborough University. In the context of MOSES project, (Ashaab 1994) developed the manufacturing model for injection moulding parts which is a common source of well-defined and structured information for supporting decision making in design for manufacture of injection moulding applications. His development provides an initial representation of process capability related to injection moulding.

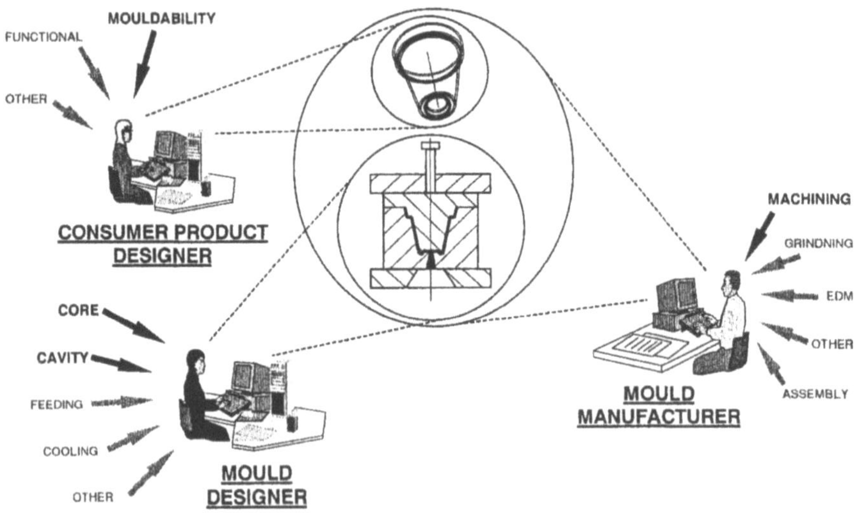

Figure 2. Multiple viewpoints in Injection Moulding.

Moreover, in the MOSES project, (Lee 1996) made an initial exploration with the definition of design for manufacture applications using the Manufacture Model in the Product Model environment. In his development he identified the problem with interactions between mouldability view and core and cavity views in design for manufacture injection moulding. His research was concerned with simple geometric relationships to specify interactions and has only a geometric representation in the product model. The work developed by (Lee 1996) has been used as a key start point to this research. In addition (Lee and Young 1998) have investigated how design for manufacture information can be provided back to design as concurrently as possible. They have identified the necessity to create mechanisms for translating information between domains in a concurrent engineering environment.

The research reported in paper has focused on three areas related to injection moulding: the design of a plastic part such as a consumer product, the design of a mould, and the manufacture of the mould to make the consumer product. In these areas there are a number of specific views to be addressed as illustrated in *Figure 2*. The work reported here concentrates on the interactions between four of them: Mouldability from Consumer product design; Core and Cavity from Mould Design and Machining from Mould Manufacture. The next sections highlight the issues addressed in this research relating two fields identified in section in this section i.e. the structure of information in the product model needed to support different but related views of the product and the definition of the relationships between these sets of information.

2.2 Product data model

To deal with multiple views in an injection moulding environment it is necessary to understand that each specific application has to hold an information structure within the product model which is able to support the function of the application. In the Product Data Model there is a need to define data structures, which can support all applications. The product data model representation has been explored for two reasons. Firstly, so it can provide support for its individual area of application and secondly, so it can support the movement of information from one application area to another. The approach taken to explore the product data model structure is not the main focus of this paper.

2.3 Defining information relationships

Given that, considered that the data structures have been defined for individual views correctly as highlighted in section 2.1, the remaining challenge is to understand how the information captured in one view can be shaped in a meaningful form for another view during the information transfer process. This research, which has been reported in this article, has explored the understanding of the way that information can be translated between views as also illustrated in *Figure 3*. This results in an exploration of Translation Mechanisms, which will be discussed more over.

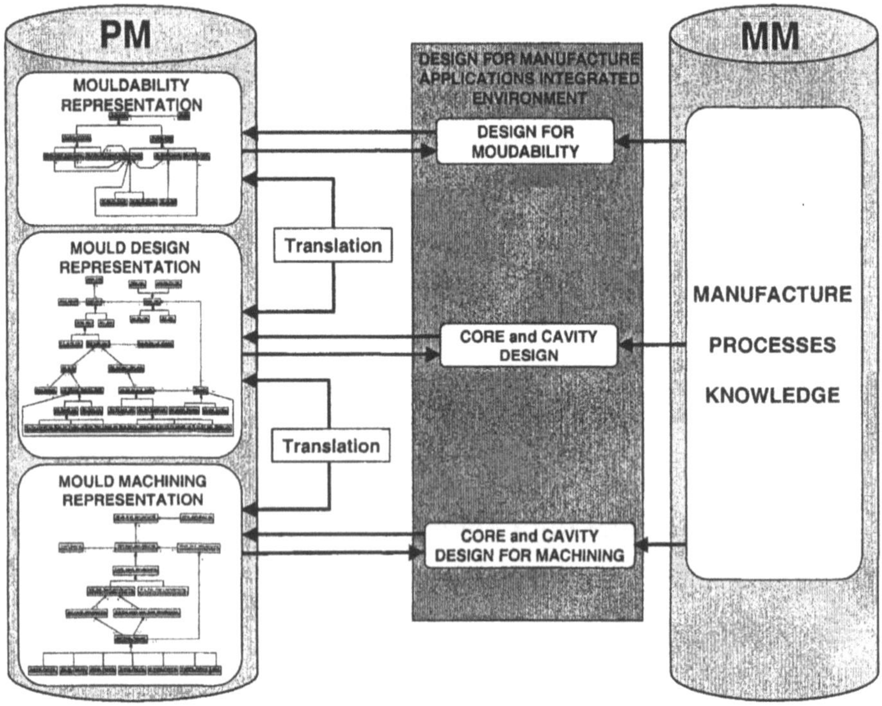

Figure 3. Translation needed to support multiple view approach.

2.4 Translating information from mouldability view to the mould design view

This section describes the research performed to define the translation mechanisms that convert information from one view to another. In this research two areas of translation have been explored. The first is the translation mechanisms for converting information from the mouldability view into cavity and core design in the mould design view. The second is the translation mechanisms for converting information from the cavity and core insert design into cavity and core insert manufacturing information in the mould manufacturing view. The mouldability, mould design and mould manufacturing views. Section 2.4.1 conceptually defines the translation mechanisms for translating information between mouldability and mould design views. Section 2.4.2 defines the translation mechanism for converting information between mould design and mould manufacturing views. In both sections the methods used for translating information from one view to another are defined and the translation mechanisms are presented using UML (Unified Modelling Language) language.

2.4.1 Critical aspects of information translation between mouldability and mould design views

There are several aspects to be discussed in terms of translating information between the mouldability view and the cavity and core design in the mould design view. This section discusses the importance of them in the context of what information contained in the mouldability view, needs to be translated for the cavity and core design in the mould design.

Figure 4 highlights the critical information that is necessary for translating from mouldability to cavity and core design. In terms of cavity design translation the critical information is the geometry of the cavity and gate features and their attributes whilst for core design the critical information includes core geometry and ejection features.

The geometric information of the cavity and the core is as it is on geometric shape of the plastic part and the position of the parting line of the mould. Also, the geometry of the cavity and the core are related directly with the shrinkage of the plastic material of the part. The gate and ejection features are also critical because they will affect the mouldability of the plastic part. The following sections will explore each particular critical aspect of translation information between mouldability and cavity and core design in the mould design views. In addition the transformation of the product geometry to accommodate effects of material shrinkage is assumed to have taken place during the design phase. In translating between views the appropriate transformed geometry is extracted from redesigned product in the database.

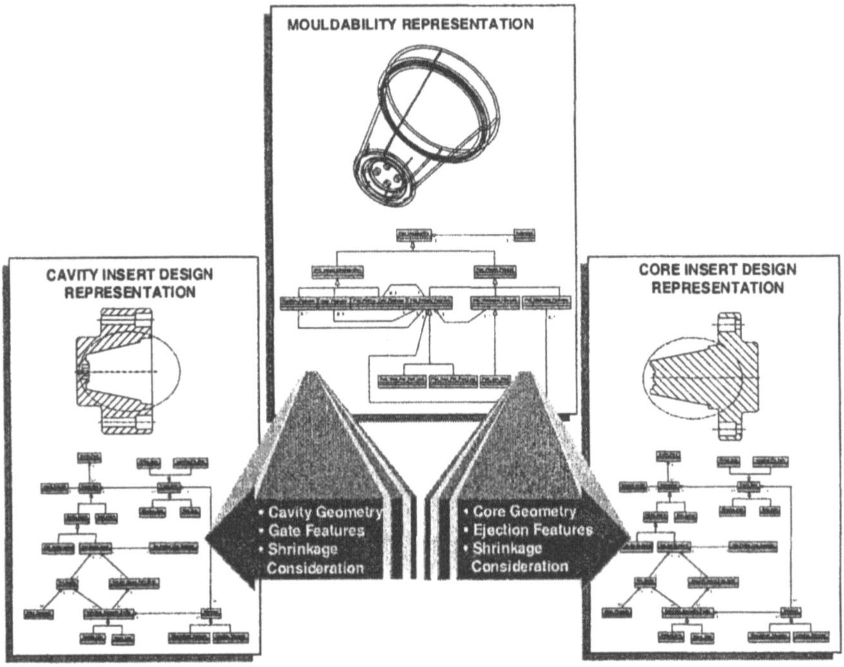

Figure 4. Information to be translated from mouldability view to cavity and core design in the mould design view

2.4.2 Mouldability primary features to cavity design

This section explores the mouldability information contained in the primary features and how this is translated to cavity design features in order to generate the main shape of the cavity for the rotational and prismatic walled products. For rotational products the main profile of the cavity is generated by extracting the external profile of each rotational mouldability primary feature that is located below the parting line plane. *Figure 5* illustrates how an external profile represented by the rotational mouldability primary features class can be translated as an individual geometric profile (*Cav_Taper_Straight_line*), which is aggregated by the *Rotational Cavity* class. The figure shows the external geometric profile information of a rotational wall being re-dimensioned according to the shrinkage coefficient of the material and being translated in terms of the cavity main profile for the cavity insert design. The points P1 and P2, and an axis that aggregates the start point and endpoint represent the geometric information of the external profile before considering shrinkage of the material. This profile is illustrated in the figure in grey. The profile illustrated by the point P1' and

P2', and an axis that aggregates the start point' and endpoint' represent the new geometric definition of the profile which belong to the cavity design in the mould design view after shrinkage consideration. This profile is illustrated in the figure in black colour.

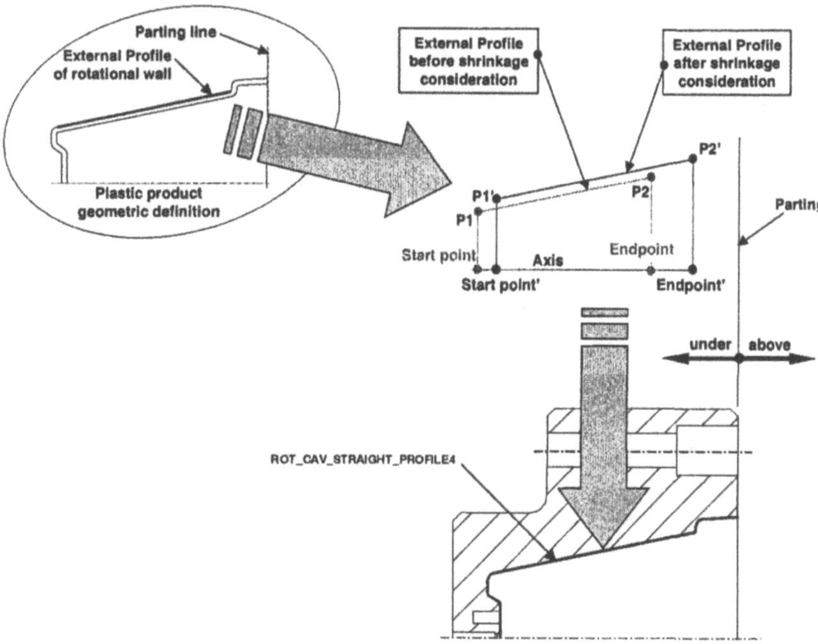

Figure 5. Example of translation of a rotational mouldability primary feature (rotational wall) into rotational cavity design.

The geometric translation of the rotational walls in terms of cavity design is shown in *Figure 6*. The figure describes the criteria that the translation mechanisms use for translating the geometric information contained in the class *Rotational Mouldability Primary Feature* (*Rot_Primary_Features*). The figure "a" shows that "Dia_ext_1" is greater than the value of the "Dia_ext_2" the translation mechanism will get the points P1' and P2', otherwise it will get the points P1 and P2 respectively. Also, the figure "b" shows the criteria used for the translation mechanisms for converting a rotational wall parallel to the parting line into *Cavity Straight Line* profile.

(a) Translation of the geometry of a rotational wall perpendicular to the parting line

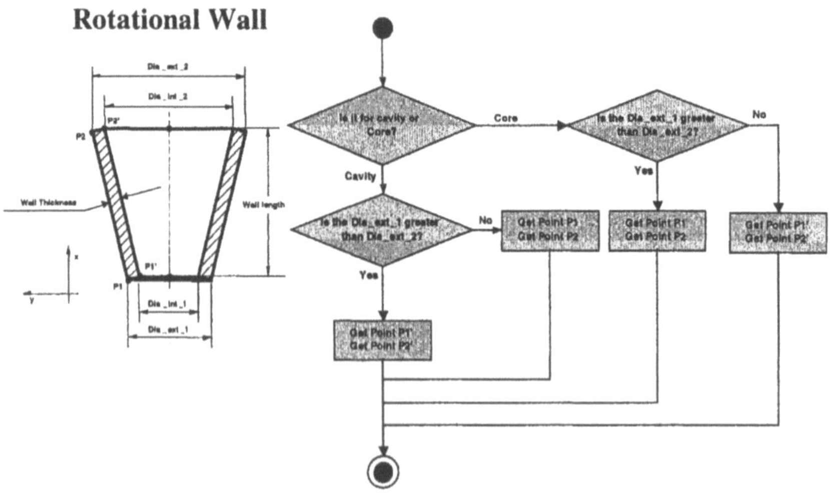

(b) Translation of the geometry of a rotational wall parallel to the parting line

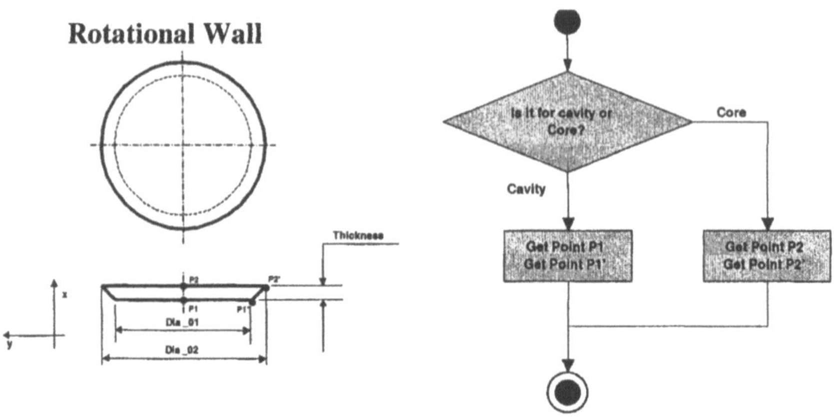

Figure 6. Geometric translation of the Rotational Mouldability Primary Features.

Similarly, to generate the main shape of the prismatic cavity it is necessary to extract the external surface of each prismatic mouldability primary feature that is located under the parting line plane. *Figure 7* illustrates the external plane surface of the *Prismatic Mouldability Primary Features* class being translated as an individual geometric surface

(*Cav_Per_Plane_Surf*) of the *Prismatic Cavity* class. The figure explores the external geometric surface of a prismatic wall being re-dimensioned according to the shrinkage coefficient of the material and being translated in terms of cavity shape in the cavity insert. The surface highlighted in the figure aggregates four points P1, P2, P3 and P4 before the shrinkage of material has been considered. After the shrinkage consideration the surface increased its size and translating the points P1, P2, P3 and P4 into P1', P2', P3' and P4' respectively.

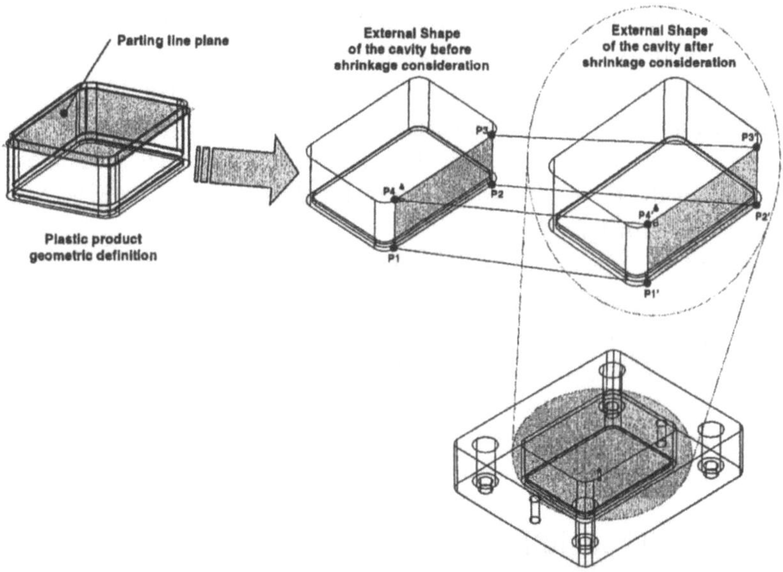

Figure 7. Example of translation of a prismatic mouldability primary feature (plane wall) into prismatic cavity design.

As described previously, each primary mouldability feature must be translated in terms of cavity design. After the plastic product has been resized, it is necessary to convert each feature into cavity and core design features in the mould design view. For rotational products it is necessary to extract geometric information from the external profile of each rotational primary and transition feature in order to generate the main profile of the cavity with its attributes and associated objects.

The steps of the translation process for modelling the rotational cavity design representation from the rotational plastic product information contained in the mouldability are:

i) to search the wall mouldability information;
ii) to get the specific external profile with its associated objects and store the information in the right form in the database;
iii) to match the information contained in the extracted profiles regarding with the respective cavity profile that is defined in the mould design view.

These steps are illustrated in *Figure 5* by considering the external profile of the wall and examining the co-ordinate values of the point P1(y) and P2(y), defined by two distinct points P1 and P2. In this case this profile will be translated as a *Cavity Taper Straight Line* (*Cav_Taper_Straight_ Line*) in the cavity design view. However if the value of the P1(y) has the same value of the value of the P2(y), it has to be translated as a *Cavity Parallel Straight Line* (*Cav_Par_Straight_Line*).

In both of the above cases the values of the co-ordinates of the points P1(x) and P2(x) must be different. The same approach can be used for defining the *Rotational Wall Parallel to the Part Line* (*Rot_Wall_Par_Part_Line*) translation mechanism. But in this particular case the value of the co-ordinate of the points P1(x) and P2(x) are the same and P1(y) and P2(y) are different. For this reason the external profile will be translated as *Cavity Perpendicular Straight Line* (*Cav_Per_Straight_Line*).

2.5 Translating information from mould design view to the mould manufacturing view

To translate information between core insert design view and core insert manufacturing view a several points must be considered. This section discusses the importance of them in the context of what information is contained in the core insert design view, which needs be translated in terms of core insert manufacturing in the mould manufacturing view. For this case the authors assume that the core insert has been designed correctly and stored in the product model. *Figure 8* illustrates the critical information that must be translated from core design to core manufacturing views. The figure is highlighting the information that must be translated between core insert design to machining in the core insert manufacturing view.

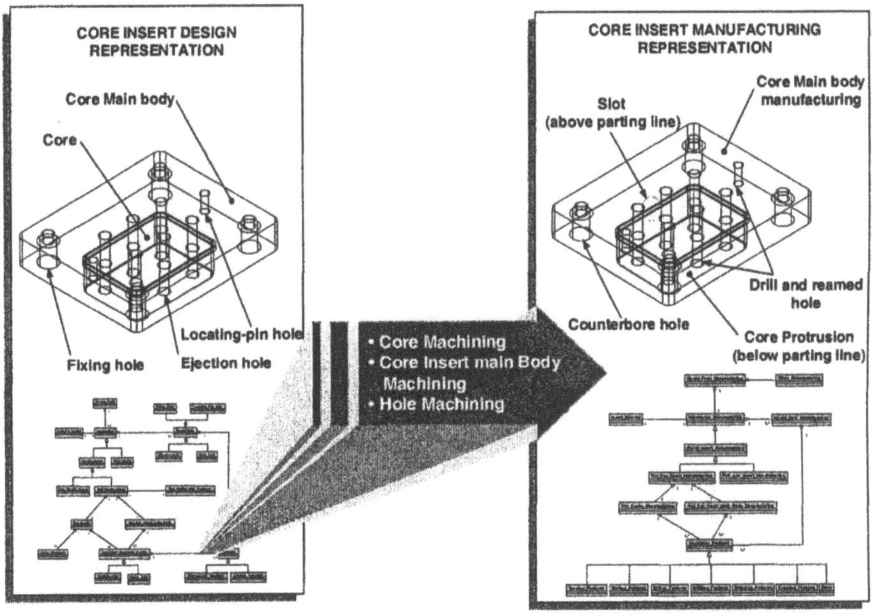

Figure 8. Information to be translated from core insert design view to core insert manufacturing view.

In order to translate the information contained in the *Rotational Cavity* class to the *Rotational Cavity Manufacturing* class, using a rotational cavity as an example, the *Cavity Individual Geometric Profile* class *(Cav_Individual_Geometric_Profile)* must be converted into a *Machining Feature* class (*Machining_Features*). The *Individual Geometric Profile* has a geometric definition and is associated with the tolerance and the material of the insert. The tolerance information is contained in the class tolerance *(Tolerance)* and the material of the cavity insert is in the class insert material *(Insert_Material). Figure 9* illustrates the information that is contained in the *Cavity Individual Geometry Profile* class (*Cav_Individual_Geometric_Profile*) with its associated tolerance and insert material classes being translated as *Machining Features* (*Machining_Features*). This machining feature is associated with the *Rotational Cavity* class *(Rot_Cavity)*, with the class *Tolerance (Tolerance)* and the *Insert Material* class *(Insert_Material)*, being translated as *Facing Boring* machining feature (*Facing_Boring*). This translation can be made by considering all the profiles that compound the cavity profile together or by considering each profile individually. The example shown in the figure considers an individual translation.

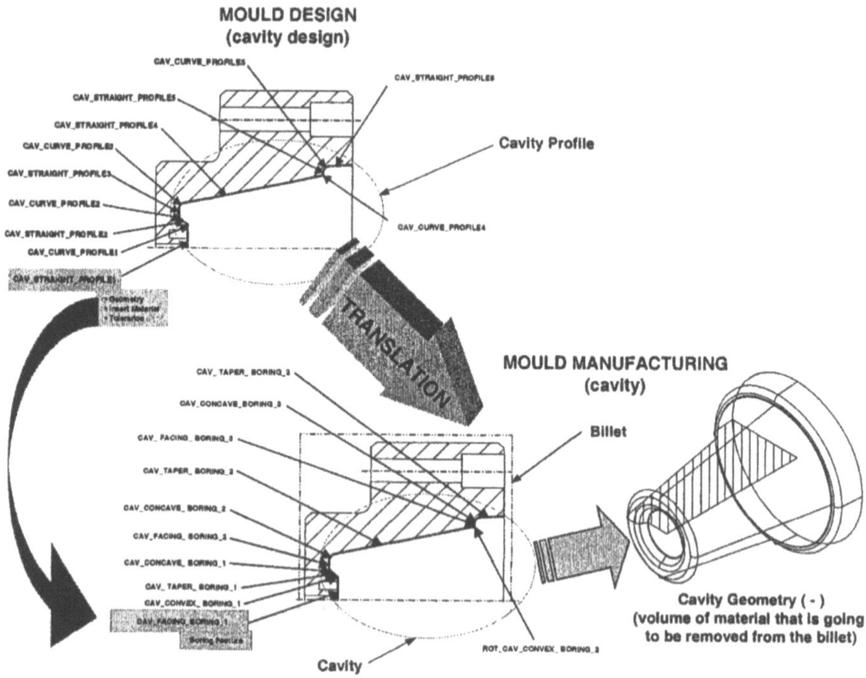

Figure 9. Example of translation of a rotational cavity into a rotational cavity manufacturing.

For a rotational cavity each *Cavity Straight Line* profile and each *Cavity Curve Line* profile must be translated in terms of *Machining Features* (*Boring_Features*) in order to generate the machining features which are going to machine the cavity. The *Cavity Straight Line* profiles can be parallels, perpendiculars or tapers to the parting line while the *Cavity Curve Line* can only be a curve. Whenever the Cavity Straight Line is parallel to the parting line (*Cav_Par_Straight_Line*), it will be translated as machining feature *Facing Boring* (*Facing_Boring*) due to the profile being perpendicular to the axis of revolution of the feature. Whenever the *Cavity Straight Line* is perpendicular to the parting line (*Cav_Per_Straight_Line*), it will be translated as a machining feature *Horizontal Boring* (*Horizontal_* Boring) because the profile in this case is perpendicular to the axis of revolution of the feature. If the *Cavity Straight Line* is tapered to the parting line (*Cav_Taper_Straight_Line*), it will be translated as machining feature *Taper Boring* (*Taper_Boring*). The definition of the translation mechanism of the *Cavity Straight Line* is illustrated in *Figure 10a.*

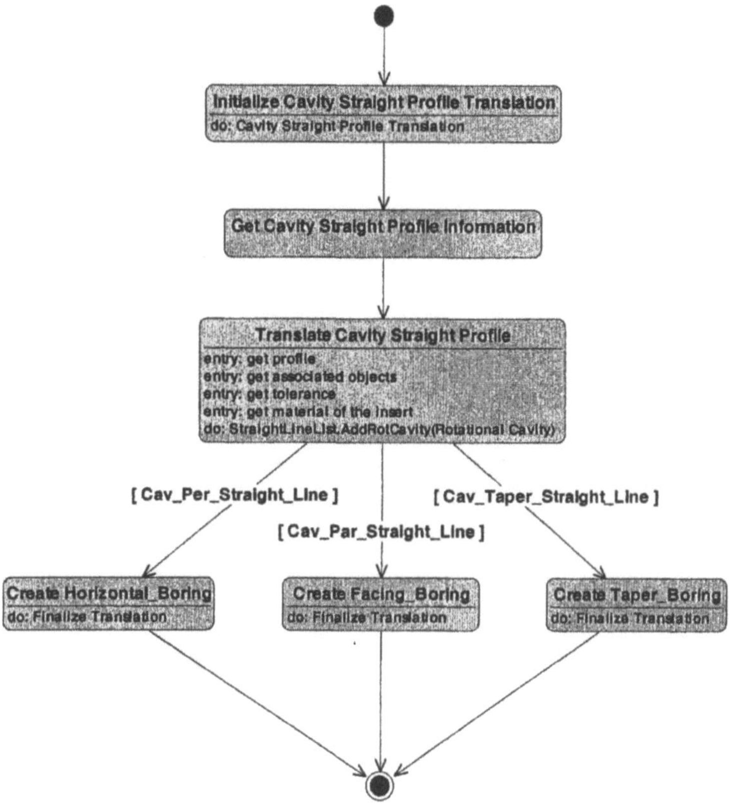

Figure 10a. The definition of the translation mechanism of the Cavity Straight Line

The translation mechanism will initially search the straight profiles in the product model. It will get the profile, the associated objects, tolerance and the material of the insert. Then it will translate each cavity straight profile into facing, horizontal or taper boring using the criteria described at the beginning of this paragraph. To translate cavity curve lines into machining features, it is necessary to convert each *Cavity Curve Line* (*Cav_Curve_Line*) into *Concave* or *Convex Boring* (*Concave_Boring* and *Convex_Boring*) machining features. If the reference angle of the *Cavity Curve Line* is bigger than 90 and less than 270 degrees the translation mechanism will convert the curved line profile as *Convex Boring* (*Convex_Boring*) machining feature, otherwise, it will convert as *Concave Boring* (*Concave_Boring*) as illustrated in *Figure 10b.*

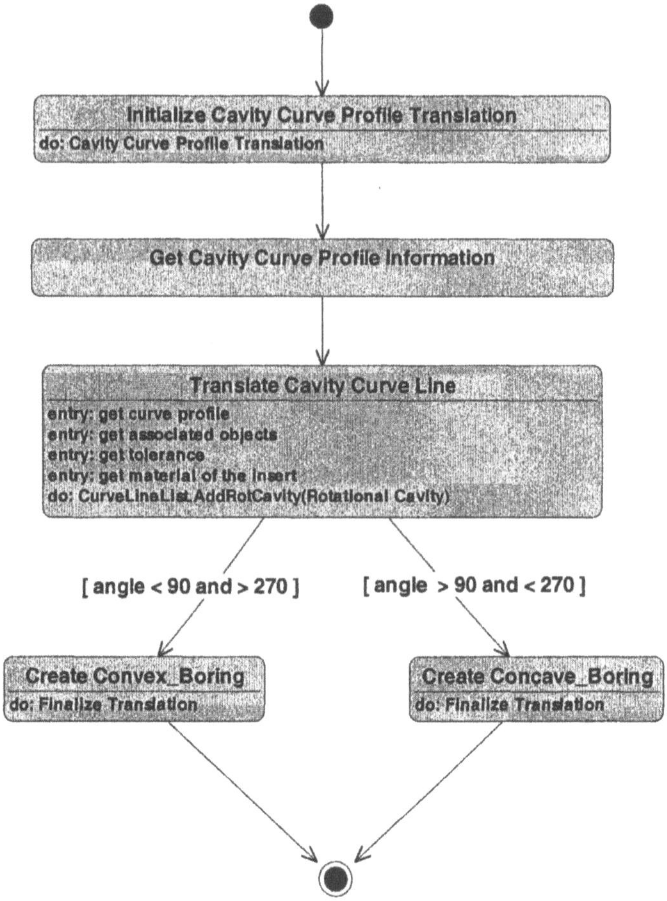

Figure 10b. The definition of the translation mechanism of the Cavity Curve Lines

The mechanism starts by searching for the cavity curve lines in the product model. It will get the curve profile, associated objects, tolerance and material of the insert.

3. CONCLUDING DISCUSSION

This paper has focused on the needs of future product design systems which must provide design teams with tools which can offer multi-viewpoint support. The advantages of utilising information modelling approaches and providing databases of information in the form of product and manufacturing models, as a precursor to multi-viewpoint decision support, have been highlighted. The research has shown that it is possible to translate information between two views once the relationship between the information between the views has been defined. Depending on the types of relationship the information will either be shared or translated from one form into another. The case of shared information is when the two views use the same piece of information without any change to it. The case of translation is necessary when one view needs the information of the another view and this information must be assigned in another format to that of the original. The translation information is more complex than shared information because the translation mechanisms must have knowledge of the relationships between the two distinct views in order to convert information from one to another. The research has shown how links can be maintained between design views of related products through the plastic part and its mould. It has also shown how links can be maintained between design and manufacture views through consideration of the mould design views and mould machining view. There is a need to extend this work to provide a more comprehensive set of both design views and manufacturing process views. For example functional design views of the plastic product, design views of the mould such as feeding, cooling, ejection and multiple manufacturing views such as EDM, grinding, assembly. The exploration of such a set of views would provide a substantial exploration of the points raised in the paragraph above concerning the likely need to extend the problem beyond a 1:1 translation. There would also be significant value in exploring the research concept against the design and manufacture of alternative, perhaps more complex, types of product other than injection moulded parts.

ACKNOWLEDGEMENTS

The authors are thankful for the financial support provided by the Brazilian Government Conselho Nacional de Desenvolvimento Científico e Tecnológico (CNPq), Pontifícia Universidade Católica do Paraná (PUCPR) and to the MIM research programme at Loughborough University with which this work was linked. The MIM research programme was funded through EPSRC grant GR/L41493.

REFERENCES

Al-Ashaab, A.H. (1994), A Manufacturing Model To Capture Injection Moulding Process Capabilities To Support Design For Manufacture, Ph.D. Thesis, Loughborough University, Department of Manufacturing Engineering.

Allada, V., Anand, S. (1995), Feature-Based Modelling Approaches For Integrated Manufacturing: State-Of-Art And Future Research Directions, Int. Journal Computer Integrated Manufacturing, Vol. 8, Number 6, pp. 411-440.

Boothroyd, G., Dewhurst, P., Knight, W. (1993), Product For Manufacture and Assembly, New York M. Dekker Inc.

Boothroyd, G. (1994) Product design for manufacture and assembly. *Computer Aided Design,* 26(7), pp. 505-520.

Chin, K.S. and Wong, T.N., Knowledge-based evaluation for the conceptual design development of injection moulding parts. *Engng Applic. Artif.Intell.,* 1996, Vol. 9, No. 4, pp. 359-376

Fauvel, O.R. (1994) Object-oriented design for manufacture. *Journal of Intelligent Manufacturing,* Vol.5, pp.1-11.

Fowler J. (1995) – STEP for Data Management, Exchange and Sharing. *Technology Appraisals Ltd.* – ISBN 1871802369.

Gupta S.K., Regli W.C., Das D. and Nau D. S. (1997) – Automated Manufacturability Analysis: A Survey. *Research in Engineering Design,* 9, pp.168-190.

Krause F.L., Ciesla M., Rieger E., Stephan M. and Ubrich A. (1995) – Features – Semantic Objects for the Integration of Tasks in the Product Development process. *Proceedings of the Computers in Engineering Database Symposium, ASME.*

Krause, F.L., Kimura, F., Kjellberg, T., Lu, S.C.Y. (1993) Product Modelling. *Annals of the CIRP, Vol. 42/2/1993.*

Lee, R.J.V, and Young, R.I.M. (1998) Information supported design for manufacture of injection-moulded rotational products. *International Journal of Production Research, Vol. 36, No 12, pp. 3347-3366.*

Lee, R.J.V. (1996), A Design For Injection Moulding Strategist In An Information Modelling Environment, Ph.D. Thesis, Loughborough University, Department of Manufacturing Engineering.

Lee, R.S., Chen, Y.M. and Lee, C.Z. (1997) Development of a concurrent mould design system: A knowledge-based approach. *Computer Integrate Manufacturing Systems,* Vol. 10, No. 4, pp. 287-307.

Pratt M. J., *"Application of feature recognition in the product life-cycle"*, International Journal of Computer Integrating Manufacturing, 6(1-2), pp. 13-19, 1993.

Rosenman, M.A. and Gero, J.S. (1996) Modelling multiple views of design objects in a collaborative design. *Computer-Aided Design,* Vol. 28, pp.193-205.

Salomons O.W., Van Houten F.J.A.M., and Kais H.J.J. (1993) *Review of Research in Feature-Based Design,* Journal of Manufacturing Systems, Volume 12/No 2.

Salomons, O.W., Houten, F.J.A.M.V., Kais, H.J.J. (1993), Review Of Research In Feature-Based Design, Journal of Manufacturing Systems, Vol. 12, Number 2, pp. 113-132.

Sanchez, J.M., Priest, J.W. and Soto, R. (1997) Intelligent reasoning assistant for incorporating manufacturability issues into the design process. *Expert Systems with Applications,* Vol. 12, No. 1, pp. 81-87.

Willems, R., Lecluse, D., Kruth, J.P. (1995), Object Oriented Information Storage For The Design Of Injection Moulds, Knowledge Intensive CAD, Vol. 1, pp. 188-202, Proceedings of First IFIP WG 5.2, Finland.

Wooldridge, M. and Jennings, N. (1995) Intelligent agents: Theory and practice. *The Knowledge Engineering Review,* Vol. 10, number 2, pp.115-152.

A Hierarchical CAPP System Architecture for Die Cavity Machining

K. KO[1], B.H. KIM[2], D.H. KIM[3] and B.K. CHOI[1]
[1] *Department of Industrial Engineering, Korea Advanced Institute of Science and Technology, Seoul, KOREA*
[2] *Korea Institute of Industrial Technology, Seoul, KOREA*
[3] *Mold Shop, Wooyoung Co. Ltd., KOREA*

Abstract: This paper proposes a hierarchical CAPP (*computer automated process planning*) system architecture for die cavity machining. The CAPP system structure has been developed on the basis of the current practices of die cavity machining and is composed of five modules – *Form EDM, Clean-up, Finish, Semi-finish, and Roughing* planning. The individual CAPP modules are connected in a sequential manner. They are identical in structure and each module consists of four sub-functions for *feature extraction, process planning, machined surface update, and NC-code generation*. Input to the CAPP system are 1) a CAD model of the die surface geometry, 2) the geometry of the raw stock, and 3) a set of available cutting tools with machinability data. Output from the CAPP system is a set of NC-codes. A process plan example, which is generated by a prototype CAPP system developed on the proposed structure, shows that the proposed CAPP architecture is practical and valid in the die cavity machining.

Key words: CAPP, machining stage, machining features, unit machining operation (UMO), STOPP, Form-EDM, Clean-up, Finish, Semi-finish, Roughing

1. INTRODUCTION

Injection molding dies for home-electronics products as well as stamping dies for automobile panels have free-formed surface geometry with complex shapes. It usually takes tens of, sometimes hundreds of, hours to machine a die cavity from a solid block. Planning for die cavity machining needs a high level of skill and experience. Often, the time required to prepare NC tool paths for the die cavity machining takes as long as the machining time itself. There is an increasing demand for a CAPP system for the die cavity machining because of the increased pressure for shorter lead-times and higher precision, coupled with the shortage of skilled workers.

CAPP has been a hot research topic (Pratt 1984 and Kanumury 1991), and more than 150 CAPP systems have reportedly been developed (Alting 1989 and Shin 1998). However, very little is published about CAPP for the die cavity machining (Yoo 1992). (Lee *et al* 1992) investigated a rough cutting strategy for free-formed cavity machining, and (Yoo and Choi 1992) proposed a recursive CAPP logic for the machining of purely free-formed die cavities. However, neither of the papers considered the *machining features* inherent in the die cavity machining.

Presented in the paper are a hierarchical CAPP system architecture for the die cavity machining and a methodology for developing such a recursive CAPP logic in (Yoo and Choi 1992). It is based on the STOPP (sequential and tool-oriented process planning) approach proposed by (Choi *et al* 1985) and is structured according to the current practices of the die cavity machining in Korea.

To develop a hierarchical CAPP system, we have to analyse detailed machining practices and then identify rules for applying specific machining operations. In this process, the concept of machining feature plays a key role. The second section reviews the recursive CAPP logic, and the third section introduces the basic elements of the CAPP system for the die cavity machining. Based on the industrial practices of the die cavity machining, this study proposes a CAPP system structure. A process plan example, which is generated by a prototype CAPP system developed on the *Z-map* based CAM software, shows that the proposed CAPP architecture is practical and valid in the die cavity machining.

2. RECURSIVE CAPP LOGIC FOR DIE CAVITY MACHINING

A *generative* CAPP for machining involves an automatic generation of a sequence of machining operations. A purely generative CAPP employs a *backward recursive approach* (Choi 1985). In the generative CAPP, a *recursive process* automatically identifies a sequence of machining operations while recursively updating the *states* of the part. The geometry and tolerance of a part specify its state. The initial raw stock for die cavity machining is either a solid stock called *moldbase* or a *preform* that is made by rough machining, forging, or casting.

In the generative CAPP, a "planning operation" is regarded as a "metal addition" operation that is the reverse of a machining (metal removal) operation. From tool *database* (DB), each planning operation selects the cutting tools that can successfully add metal to the current machined surface to obtain a feasible set of process plans. It then selects a "best tool" according to a performance measure. Finally, it updates the *current state of the part* (that is, the machined surface is offset and the tolerance relaxed) to complete a planning step. This individual planning step is repeated until the current state of the part becomes the *final state* of the part. This planning procedure is summarized as follows (Yoo 1992):

Recursive CAPP Algorithm:

0. Input: Initial (current) state and tolerance, final state, tool DB.

1. Select all the tools that can "plan" the current state of the part.

2. Choose a tool (or a set of tools) that gives minimum cutting time.

3. Update the current state by offsetting the surface and relaxing the tolerance.

4. If the current state equals to the final state then stop, else go to 1.

When planning for sculptured surface machining, the relationships among the following three items play an important role:

- *Machining allowance*, α: the amount of intentional under-cut thickness.
- *Machining tolerance*, τ: maximum deviation of the actual machined surface from the ideal machined surface to be obtained.
- *Machining depth*, δ: the difference between the machined surface and the preform surface.

The required machining depth δ at a (machining) stage is determined as a function of the tolerance τ. Similarly, τ is also defined as a function of α. These relations are expressed as

$$\tau = f(\alpha) \quad or \quad \alpha \Rightarrow \tau$$
$$\delta = f(\tau) \quad or \quad \tau \Rightarrow \delta$$

Obviously, the machining allowance at a stage becomes the machining depth at the next stage of machining. There usually are three stages of machining in the die cavity machining: *roughing, semi-finishing,* and *finishing*. This paper assumes the following technical relationships:

$\tau_f = 0.05 \Rightarrow \delta_f = 0.3$, for finishing;

$\alpha_s\ (\equiv\delta_f) = 0.3 \Rightarrow \tau_s = \pm 0.15 \Rightarrow \delta_s = 2.0$, for semi-finishing;

$\alpha_r\ (\equiv\delta_s) = 2.0 \Rightarrow \tau_r = \pm 1.0 \Rightarrow \delta_r = \infty$, for roughing;

These data have no theoretical basis, but die makers use them as "technical standards". We will also use these rules throughout the paper.

The die cavity machining mainly uses *ball endmill (BEM)* and *flat endmill (FEM)*. The following abbreviations will be used throughout the paper:

BEM20ϕ : ball endmill of 20mm in diameter.
FEM20ϕ : flat endmill of 20mm in diameter.
Drill50ϕ : drill of 50mm in diameter.

Figure 1 shows a schematic description of the recursive CAPP logic. As depicted in *Figure 1*-a, the final geometric shape of the die cavity and its tolerance value (τ_f=*0.05mm*) are specified as an *initial state* of the part. The first step *plans* a finishing operation:

1) Select a finishing tool (*BEM20ϕ*)
2) Offset the part by δ_f (=*0.3mm*) and relax its tolerance to τ_s(=*0.15mm*).

The result is a *finishing preform* as depicted in *Figure 1*-b. The next step plans a semi-finishing operation:

1) Select a cutter (BEM30ϕ)
2) Offset the part by δ_s and relax the tolerance to τ_r(=*1mm*).

Now we have a *semi-finishing preform* to be obtained by a roughing operation as depicted in *Figure 1*-c. The roughing (i.e., pocketing) operation is planned as:

1) Select a pocketing tool FEM50ϕ.
2) Fill the die cavity up to the initial solid block ($\delta_r = \infty$).
3) Plan for a drilling operation.

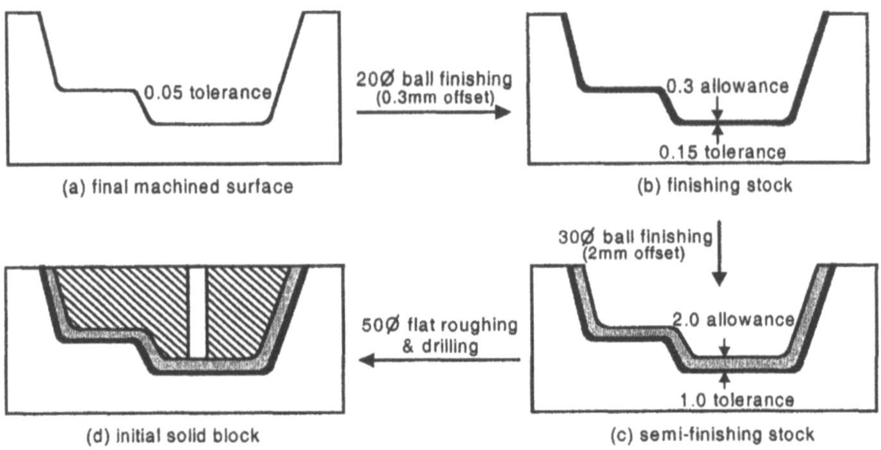

Figure 1. Recursive CAPP Logic

3. BASIC ELEMENTS OF CAPP SYSTEM

The basic elements of CAPP system for the die cavity machining are machining stages (*MS*), machining features (*MF*), and unit machining operations (*UMO*).

3.1 Machining stages

Diverse definition of machining stages has been proposed in the literature (Fallbohmer 1995, Yamazaki 1995, and Yoo 1992), with the most popular scheme being the "three-stage" scheme consisting of roughing(*R*), semi-finishing (*S*), and finishing (*F*). However, this study proposes a "five-stage" decomposition scheme based on a close examination of the die cavity machining practices employed by die-makers. *Figure 2* shows the five machining stages: roughing stage, semi-finish stage, finish stage, clean-up stage, and form-EDM stage.

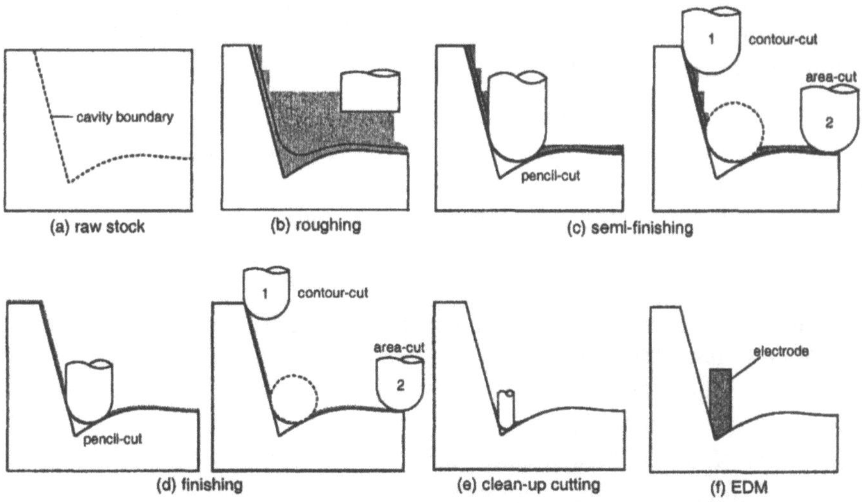

Figure 2. Die cavity machining stages

- *Roughing machining stage*: Removes metal from a raw stock as fast as possible (using heavy-duty cutters) to obtain a machined preform for the next machining stage (i.e., semi-finishing).
- *Semi-finish machining stage*: Obtains a "finishing" preform having a uniform thickness of the finishing allowance.
- *Finish machining stage*: Obtains the final shape of the die surface with the required surface-finishes. To have a shortened machining time, this stage usually uses relatively large cutters (ball endmills of 20-30mm diameter).

▪*Clean-up machining stage*: Using smaller ball endmills (and flat endmills), removes uncut volumes at the concave regions where the "finishing" ball endmills could not reach.

▪*Form-EDM machining stage*: Using special form-cutters or EDM-electrodes, cuts the sharp concave edges and the special depression that can not be produced with a ball endmill (The EDM-electrodes are usually machined by ball endmills).

3.2 Machining features

Most of the free-form features introduced in the literature (Park 2000, Shin 1998, and Storr 1995) are concerned with machining applications. This study assumes that the die surface is visible from above (i.e., no "negative draft" in the die surface) and that NC machines can perform all 3-axis machining operations. Based on a close observation of the current die cavity machining, this study takes into account five types of machining features: wall type machining features, floor type machining features, concave-strip type machining features, volume type machining features, and prismatic type machining features (Choi 1996).

1) Wall type machining features

▪*Wall*: a near vertical surface with positive draft angles (say, less than 30 degree) (*Figure 3*-a).

▪*Vertical wall*: a wall with a zero draft-angle (*Figure 3*-b).

▪*Cavity wall*: the entire wall of a die cavity forming a closed strip of walls (*Figure 3*-e)

▪*Core wall*: the entire wall of a core or a protrusion (*Figure 3*-f).

2) Floor type machining features

▪*Floor*: a near horizontal surface with an inclination angle (say, less than 30 degree) (*Figure 3*-a, b, and c)

▪*Compound floor*: a "compound" area in the die surface containing a number of floors and other features (*Figure 3*-d).

3) Concave-strip type machining features

▪*Vertical concave fillet*: a concave blending surface between two draft(or vertical) surfaces, or a concave edge(a round with a very small fillet radius or a sharp concave edge)(*Figure 3*-a).

▪*Sloped concave fillet*: a concave blending surface between a slant and a draft surface, or a sloped concave edge (*Figure 3*-c).

▪*Horizontal concave fillet*: a concave blending surface between floor and draft surfaces, concave edge, or a round groove (*Figure 3*-b).

4) Volume type machining features (*Figure 3*-h)

▪*Cavity volume*: a volume in the die cavity that is defined as a Boolean difference between a preform and a machined surface.

▪*Shoulder volume*: an "open" volume of metal to be removed.

5) Prismatic type machining features

▪*Step*: a stepped feature consisting of a vertical plane and a horizontal plane.

▪*Slot*: a rectangular slot.

▪And more types: there are many papers dealing with prismatic-type features (Han 1998, Lee 1998, and Xu 1998).

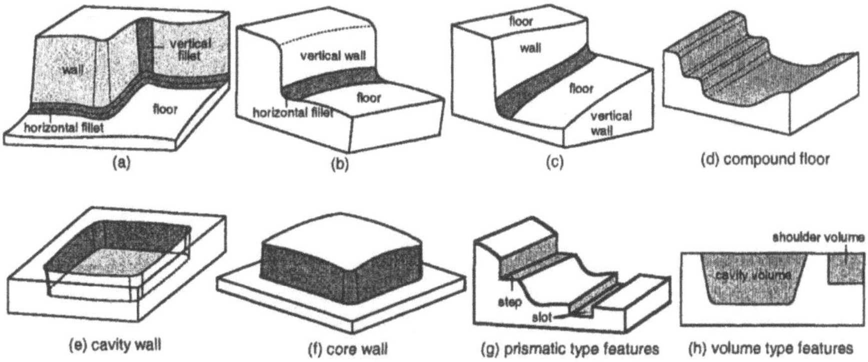

Figure 3. Die cavity machining features

3.3 Unit machining operations

A unit machining operation (UMO) is defined as a specific pattern of the die cavity machining operation performed using a single cutter (Choi 1998). Introduced in the study are a total of 17 UMOs that are widely employed in the die cavity machining. Based on the mapping relationship between the machining features and UMOs, this study groups the UMOs into the following five types: wall type UMOs, floor type UMOs, strip type UMOs, volume type UMOs, and special UMOs (*Figure 4* shows some UMOs usually used in the die cavity machining).

1) *Wall type UMOs*: z-constant contour-cut, BC(boundary curve)-parallel contour-cut, helical contour-cut.
2) *Floor type UMOs*: XY-parallel area-cut, BC-parallel area-cut, BC-normal area-cut, CPO(contour parallel offset) area-cut, and spiral area-cut.
3) *Strip type UMOs*: pencil-cut, strip-parallel fillet-cut, and strip-normal fillet-cut.
4) *Volume type UMOs*: drilling, pocketing, and shouldering.
5) *Special UMOs*: form-cutting (slotting and 2D-contouring) and EDM.

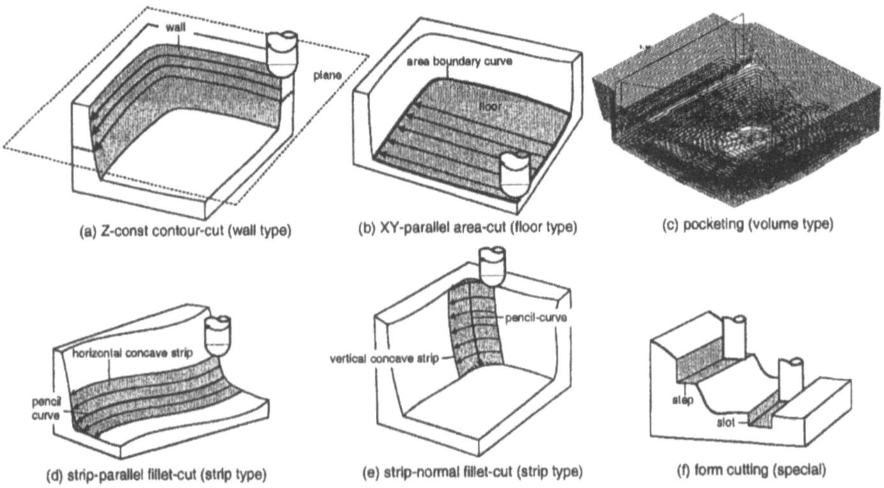

Figure 4. Some UMOs used in the die cavity machining.

4. CAPP SYSTEM ARCHITECTURE FOR DIE CAVITY MACHINING

The proposed CAPP system architecture is built around the machining practices. By reversing the sequence of the machining stages introduced in the previous section, this study constructs a five-stage CAPP logic as follows:

Five-Stage CAPP Logic

0. Input: Raw Stock; Die Cavity; Tool List;
1. Plan for the Form-EDM machining;
2. Plan for the Clean-up machining;
3. Plan for the Finish machining;
4. Plan for the Semi-finish machining;
5. Plan for the Roughing machining;
6. Output: NC-codes.

Figure 5 shows the overall structure of the proposed CAPP system, which is based on the five-stage CAPP logic. The individual CAPP modules are connected in a sequential manner and the planning operations are driven by available cutting tools. However, we characterize the proposed CAPP structure as "sequential" and "tool-oriented" (instead of "recursive" and "generative"). Specifically, it is an adaptation of the *STOPP* (Sequential and Tool-Oriented Process Planning) approach proposed by (Choi *et al* 1985).

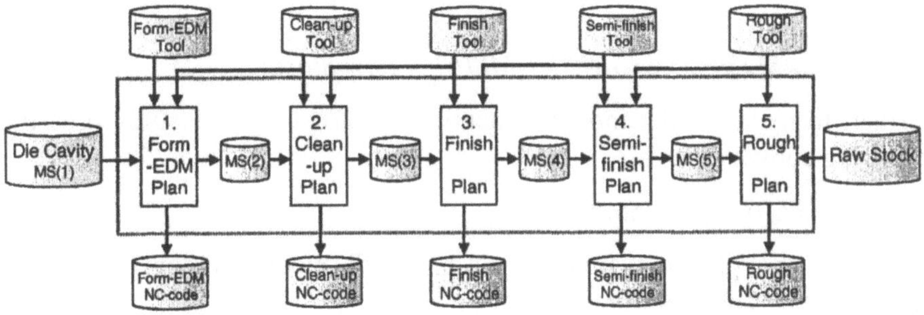

Figure 5. Overall Structure of the CAPP System

Figure 6 shows the internal structure of each CAPP module that includes the following objects:

- *MS[j]: machined surface at the planning stage j.*
- *TL[j]: tool-list at the planning stage j.*
- *MR[j]: machining rules to be used at the planning stage j.*
- *FL[j]: feature-list for the features extracted at the planning stage j.*
- *PP[j]: process plans obtained at the planning stage j.*
- *NC[j]: NC-codes generated at the planning stage j.*

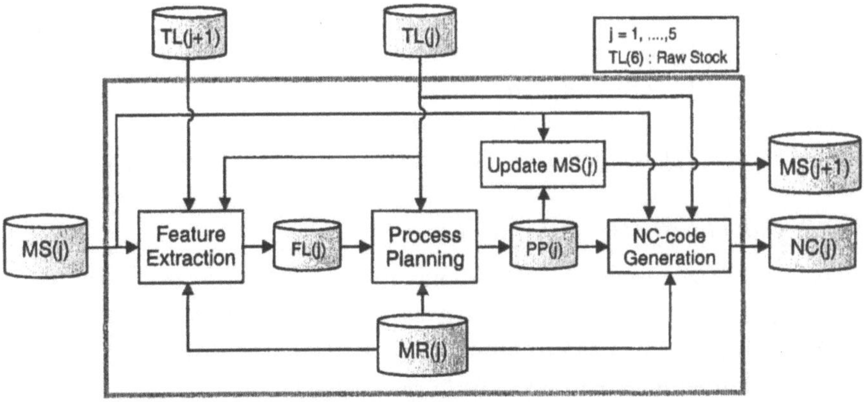

Figure 6. Structure of the local CAPP Module

Each CAPP module has its own set of built-in machining rules, *MR[j]* (Machining rules indicate how to cut for machining features extracted from the postform at each machining stage), and its input and output are as follows:

- *Input: MS[j], TL[j], TL[j+1].*
- *Output: NC[j], MS[j+1].*

Note that

$$MS[1] \equiv Die\ Cavity; \quad (MS[6] \equiv Raw\ Stock)$$

Further, to make all the modules identical in structure, this study uses the following convention:

$$TL[6] \equiv Raw\ Stock.$$

Internally, the planning function of a CAPP module j consists of the following four sub-functions:

1) Extract a set of features *FL[j]* from *MS[j]* using *MR[j], TL[j]* and *TL[j+1]*.
2) Plan for machining processes *PP[j]* from *FL[j]* using *MR[j]* and *TL[j]*.
3) Update the machined surface from *MS[j]* to *MS[j+1]* using *PP[j]*.
4) Generate NC-codes *NC[j]* from *MS[j]* using *PP[j], TL[j]*, and *MR[j]*.

Thus, this study formalizes the "Five-stage CAPP logic" in terms of the following "sequential and tool-oriented process planning" (STOPP) algorithm:

STOPP Algorithm:

```
0. Input : MS[1] = Die Cavity; TL[6] = Raw-Stock;
          { TL[j], j = 1,...,5} = Tool-Lists;
1. For j = 1 to 5 do {
       FL[j] ⇐ extract_feature(MS[j] | MR[j], TL[j], TL[j+1]);
       PP[j] ⇐ plan_processes(FL[j] | MR[j], TL[j]);
       MS[j+1] ⇐ update_MS(MS[j] | PP[j]);
       NC[j] ⇐ generate_NC(MS[j] | PP[j], TL[j], MR[j]);
       }
2. Output { NC[j], j = 1,...,5 };
```

The machined surface *MS[j]* denotes the geometry of the die cavity at stage j (The technological data such as uncut allowance, tolerance, and cutting depth are treated as machining rules *MR[j]*). The feature list *FL[j]* is a set of machining features F_i, where F_i is defined as follows:

$F_i = feature\_name[parameter \mid update];$

where, parameter = {α *(draft-angle)*, ρ *(radius)*, δ *(depth)*, ω *(width)*, *etc.*};

update ={ ρ *(preform fillet radius)*, o *(offset distance)*, *etc.*}

The "update" field contains methods of updating machining features (or the postform-surface). If an argument is omitted, its value is assumed to be zero. Actually, each planning stage takes into account only a limited number of types of feature elements like the following (finishing stage) example:

$FL[3] = \{ concave\_fillet[\ \mid o=\delta_f, \rho_p = R_s], wall[\mid o=\delta_f], floor[\ \mid o=\delta_f]\}$

where, Subscripts f, s denote finish, semi-finish, respectively;

R: minimum cutter radius;
δ_f*: machining depths at finish stage.*

For the sake of simplicity, this study excludes *convex features* (Specifically, the *convex-strip features* are treated as a part of *floor* or *wall*).

A process-plan *PP[j]* contains a number of machining-plans $\{P_i\}$. A machining-plan consists of a machining-features(MF) and a UMO, and is expressed as follows:

$$P_i \equiv (MF_i,\ UMO_i\ ).$$

Finally, the update operation is expressed as

$MS[j+1] \Leftarrow update\_MS(MS[j] \mid type(MF),\ v);$ *v: real number.*

where, *type* = (*offset, fillet, fill*).

The real number *v* denotes the offset distance, fillet radius, or the amount of fill. If the cavity is filled up to the rim, it is expressed as

$$MS[j+1] \Leftarrow update\_MS(MS[j] \mid fill(F1),\ \infty);$$

5. ILLUSTRATIVE EXAMPLE

The STOPP algorithm for the die cavity machining requires five CAPP modules (*Figure 5*), and each module consists of four sub-functions (*Figure 6*). We define the mapping relationships between machining features and UMOs under the proposed CAPP structure, and develop a prototype CAPP system on the *Z-map* based CAM software called *Soft-Master®*.

The prototype CAPP system generates a sequence of machining operations for an example of the die cavity machining. The input data of the example is a CAD model of a (17 inches) monitor front-mask consisting of 83 trimmed surfaces and a raw stock having the size of 545mm length, 504mm width, and 100mm height.

▪*MS[1]* = Die cavity (*Figure 7*-a shows the CAD and the Z-map model of the example).

▪*TL[6]* = Raw stock (moldbase, *Figure 7*-b).

▪*Tool lists:*

TL[1] = { FEM20ϕ }
TL[2] = { BEM4ϕ}, BEM8ϕ, BEM12ϕ, BEM16ϕ }
TL[3] = { BEM20ϕ}
TL[4] = { BEM30ϕ}
TL[5] = { FEM30ϕ, Drill25ϕ, Drill50ϕ }

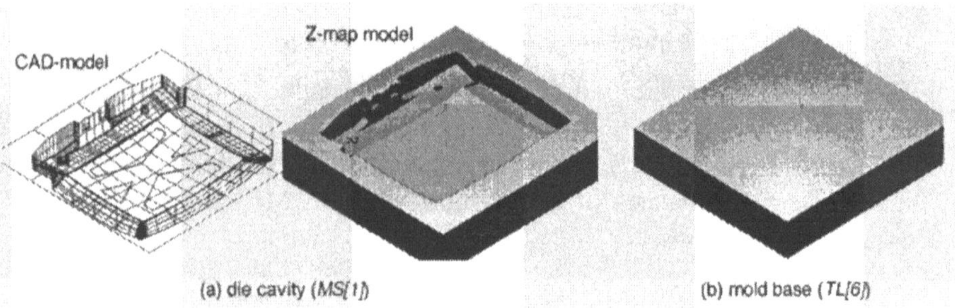

Figure 7. Input data of an example process planning

Stage 1: Form-EDM Plan

0) Machining rule, *MR[1] = {R1}*: machining rules for *Form-EDM* planning.
 ▪R1 = Form features (step) are machined with FEM20ϕ.

1) Feature extraction, *FL[1] = {F1} ⇐ extract_features (MS[1]|MR[1], TL[1], TL[2])*: Form features with fillet radius ρ less than 2mm (BEM4ϕ) are extracted.
 ▪$F1 = step[\ |\ \rho_p = 2.0]$.

2) Process plan, *PP[1] = {P1} ⇐ plan_process (FL[1]|MR[1], TL[1])*
 ▪*P1 ≡ (F1, UMO1) = (step, 2D-contouring[FEM20ϕ]]*.

3) Machined surface, *MS[2] ⇐ update_MS (MS[1]|PP[1])*
 ▪Call "$MS[2] = update\_MS(MS[1]\ |\ fillet(F1, \rho_p = 2.0)$" to remove F1 by filleting it with a fillet-radius of 2mm (making it a new fillet of radius 2mm) (*Figure 8*-b).

4) NC-code, *NC[1] ⇐ generate_NC (MS[1]|PP[1], TL[1], MR[1])*
 ▪Generate NC-codes for the form cutting operation P1 with the FEM20ϕ (*Figure 8*-a).

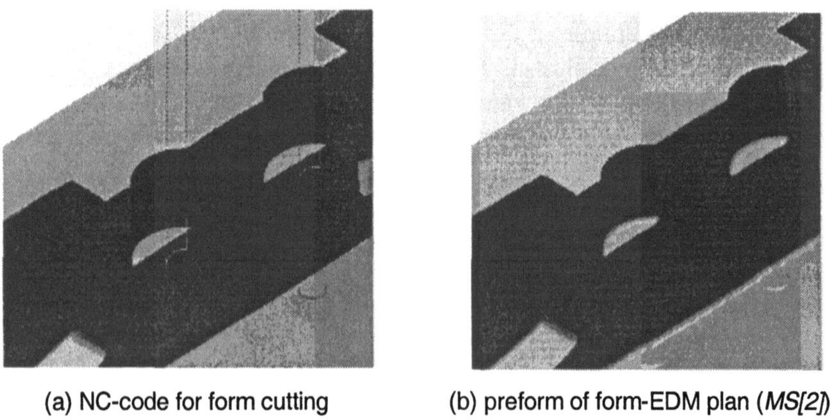

Figure 8. Form-EDM planning example

Stage 2. Clean-up Plan

0) Machining rule, *MR[2] = {R2, R3}*: machining rules for *Clean-up* planning.

▪R2 = Concave fillets with their radii less than the radius of the finish-cut tool (BEM20ϕ) are machined with a set of clean-up tools.

▪R3 = The clean-up cutting sequence is pencil-cut operations with BEM16 ϕ, BEM12 ϕ, and BEM8ϕ, and strip-parallel fillet-cut with BEM4 ϕ.

1) Feature extraction, *FL[2] = {F2}* ⇐ *extract_features (MS[2]|MR[2], TL[2], TL[3])*

▪*F2 = concave_fillet[* | ρ_p *=10.0]*.

2) Process plan, *PP[2] = {P2, P3, P4, P5}* ⇐ *plan_process (FL[2]|MR[2], TL[2])*

▪*P2* ≡*(F2, UMO2)=(F2, fillet-cut(BEM4ϕ))*.

▪*P3* ≡*(F2, UMO3)=(F2, pencil-cut(BEM8ϕ))*.

▪*P4* ≡*(F2, UMO4)=(F2, pencil-cut(BEM12ϕ))*.

▪*P5* ≡*(F2, UMO5)=(F2, pencil-cut(BEM16ϕ))*.

3) Machined surface, *MS[3]* ⇐ *update_MS (MS[2]|PP[2])*

▪*MS[3] = update_MS(MS[2]| fillet(F2),* ρ_p *=10.0)* (*Figure* 9-b).

4) NC-code, *NC[2]* ⇐*generate_NC (MS[2]|PP[2], TL[2], MR[2])*

▪Generate NC-codes for P2, P3, P4, and P5 (*Figure* 9-a).

(a) NC-codes for clean-up cutting

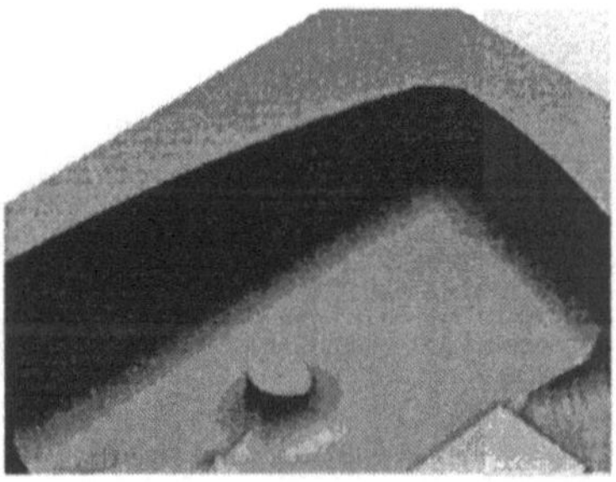

(b) preform of Clean-up plan (*MS[3]*)

Figure 9. Clean-up planning example

Stage 3. Finish Plan

0) Machining rule, *MR[3] = {R4, R5, R6, R7}*: machining rules for *Finish* planning.

▪R4 = The machining tolerance and depth for finish-cut are 0.05mm (τ_f=0.05) and 0.3mm (δ_f=0.3), respectively.

▪R5 = Concave fillets with their radii less than $\delta_f + \rho_s = 15.3$($\rho_s = 15$ is the radius of the semi-finish tool, BEM30ϕ) are "pencil-cut" machined before the main finishing.
▪R6 = The machining feature "wall" is machined by a "contour-cut" UMO.
▪R7 = The machining feature "floor" is machined by a "area-cut" UMO.

1) Feature extraction, *FL[3] = {F3, F4, F5}* ⇐ *extract_features (MS[3]|MR[3], TL[3], TL[4])*
▪*F3 = concave_fillet[| o=0.3,* ρ_p *=15.0]. // offsetting followed by filleting*
▪*F4 = wall[*α*=5°,* δ*=40 | o=0.3] : a 5°-draft wall of depth (or height) 40.*
▪*F5 = floor[| o=0.3] : offsetting*

2) Process plan, *PP[3] = {P6, P7, P8}* ⇐ *plan_process (FL[3]|MR[3], TL[3])*
▪*P6 ≡(F3, UMO6)= (F3, pencil_cut[BEM20ϕ]).*
▪*P7 ≡(F4, UMO7)= (F4, contour_cut[BEM20ϕ]).*
▪*P8 ≡(F5, UMO8)= (F5, area_cut[BEM20ϕ].*

3) Machined surface, *MS[4]* ⇐ *update_MS(MS[3]|PP[3])*
▪*MS[4] =update_MS(MS[3]| offset, 0.3).*
▪*MS[4] =update_MS(MS[4]| fillet(F3), 15.0)* (*Figure 10*-d)

4) NC-code, *NC[3]* ⇐*generate_NC (MS[3]|PP[3], TL[3], MR[3])*
▪Generate NC-codes for P6: BEM20ϕ pencil-cut (*Figure 10*-a).
▪Generate NC-codes for P7: BEM20ϕ contour-cut (*Figure 10*-b).
▪Generate NC-codes for P8: BEM20ϕ area-cut (*Figure 10*-c).

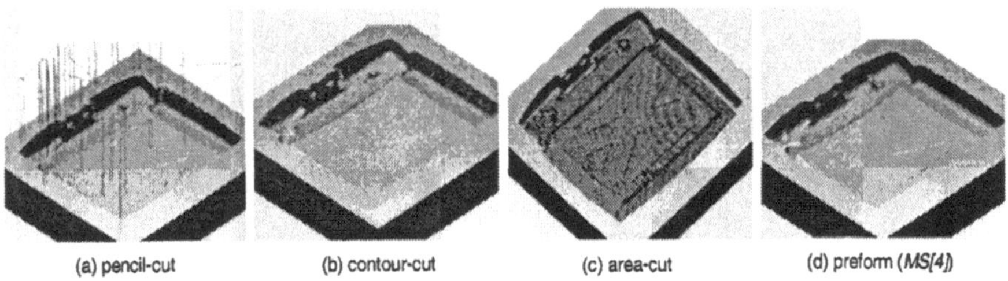

Figure 10. Finish planning example

Stage 4. Semi-finish Plan

0) Machining rule, *MR[4] = {R8, R9, R10, R11}*: machining rules for *Semi-finish* planning.
▪R8 = The machining tolerance and depth for SF-cut are 0.15mm (τ_s=0.15) and 2.0mm (δ_s=2.0), respectively.
▪R9 = Concave fillets with their radii less than $\delta_s + \rho_r = 17.0$($\rho_r = 15$ is the radius of the rough tool, FEM30ϕ) are "pencil-cut" machined before the main semi-finishing.
▪R10 = The machining feature "wall" is machined by a "contour-cut" UMO.

▪R11 = The machining feature "floor" is machined by a "area-cut" UMO.

1) Feature extraction, *FL[4] = {F6, F7, F8} ⇐ extract_features (MS[4]|MR[4], TL[4], TL[5])*

▪*F6 = concave_fillet[| o=2.0].*

▪*F7 = wall[α=5°, δ=40 | o=2.0]*

▪*F8 = floor[| o=2.0]*

2) Process plan, *PP[4] = {P9, P10, P11 } ⇐ plan_process (FL[4]|MR[4], TL[4])*

▪*P9 ≡(F6, UMO9)= (F6, pencil_cut[BEM30ϕ]).*

▪*P10 ≡(F7, UMO10)= (F7, contour_cut[BEM30ϕ]).*

▪*P11 ≡(F8, UMO11)= (F8, area_cut[BEM30ϕ]].*

3) Machined surface, *MS[5] ⇐ update_MS(MS[4]|PP[4])*

▪*MS[5] =update_MS(MS[4]| offset, 2.0).*

▪*MS[5] =update_MS(MS[4]| fillet(F6), 15.0).*

4) NC-code, *NC[4] ⇐generate_NC (MS[4]|PP[4], TL[4], MR[4])*

▪Generate NC-codes for P9: BEM30ϕ pencil-cut.

▪Generate NC-codes for P10: BEM30ϕ contour-cut.

▪Generate NC-codes for P11: BEM30ϕ area-cut.

Stage 5. Roughing plan

0) Machining rule, *MR[5] = {R12, R13, R14, R15}*: machining rules for *Roughing* planning.

▪R12 = Rough cutting is performed by pocketing, and the machining depth for the pocketing is 5mm.

▪R13 = A drill hole is needed for the pocketing, and the drill hole is drilled first by a 25ϕ drill and then by a 50ϕ drill.

▪R14 = A FEM30ϕ is used for the pocketing.

▪R15 = The tolerance of rough cutting is 1mm(τ_r=1.0).

1) Feature extraction, *FL[5] = {F9} ⇐ extract_features (MS[5]|MR[5], TL[5], TL[6])*

▪*F9 = cavity volume[δ=∞].*

2) Process plan, *PP[5] = {P12} ⇐ plan_process (FL[5]|MR[5], TL[5])*

▪*P12 ≡(F9, UMO12)=(F9, pocketing[FEM30ϕ]).*

3) Machined surface, *MS[6] ⇐ update_MS(MS[5]|PP[5])*

▪*MS[5] = update_MS(MS[5] | fill(F9), ∞)*

4) NC-code, *NC[5] ⇐generate_NC (MS[5]|PP[5], TL[5], MR[5])*

▪Generate NC-codes for P12: drilling with Drill25ϕ and Drill50ϕ, and pocketing with FEM30ϕ.

6. CONCLUSIONS AND DISCUSSIONS

This paper presents a sequential and tool-oriented CAPP structure for die cavity machining. The proposed CAPP system architecture is a practical one because it is based on industrial practices in die cavity machining. In other words, if a company prefers different rules, the CAPP system easily incorporates them. We developed a prototype CAPP system on the Z-map based CAM software. Preliminary results showed that the proposed CAPP system architecture is practical and valid in die cavity machining.

The novel contribution of the study is more in the formulation of CAPP problem (than in solving the problem). The authors believe that the proposed CAPP system architecture needs more testing and improvement. We then hope that the current contribution will encourage more researchers in the CAPP area to work on the process planning problem in die cavity machining.

The proposed CAPP architecture can be extended to the machining of negative draft areas, which are not machined by 3-axis machining operations. Specifically, if the negative draft areas are extracted from the die surface and a feasible tool orientation is determined, the proposed architecture is applied to the machining of the negative draft areas in the same manner.

ACKNOWLEDGEMENT

The Ministry of Science and Technology of Korea supported this research through a "*National Research Lab Project*" granted to Prof. Choi.

REFERENCES

Alting, L. and Zhang H. (1989), Computer Aided Process Planning: the state-of-art survey, *Int. Jr. Prod. Res.*, Vol. 27, No. 4, pp. 553-585

Choi, B. K. and Barash, M. M.(1985), STOPP: an approach to CADCAM integration, *CAD*, Vol. 17, No. 4, pp.162-168

Choi, B. K. (1991), *Surface Modeling for CAD/CAM*, Elsevier

Choi, B. K., Kim, B. H., and Kim, D. H. (May 1996), Feature Extraction for CAPP of Die Cavity Machining, *IFIP WG5.2 workshop on Geometric Modeling in CAD*, Airlie Center, VA, U.S.A.

Choi B. K. and Jerard R. B. (1998), *Sculptured Surface Machining*, Kluwer Academic Publishers

Fallbohmer, P, Akgerman, N and Altan, T, Survey of the U.S. Die and Mold Manufacturing Industry, *ERC/NSM-D-95-41 (1995), ERC for Net Shape Manufacturing*, The Ohio State University, U.S.A.

Han, J. H. and Rosen, D. (1998), Special panel session for feature recognition at the 1997 ASME computers in engineering conference, *CAD*, Vol. 30, No. 13, pp.979-982

Kanumury, M. and Chang, T. C. (1991), Process planning in an automated manufacturing environment, *Jr. of Manufacturing System*, Vol. 10, No. 1, pp.67-78

Lee, J. Y. and Kim, K. (1998), A feature-based approach to extracting machining features, *CAD*, Vol. 30, No. 13, pp.1019-1035

Lee, Y. S. Choi B. K. and Chang T. C. (1992), Cut distribution and cutter selection for sculptured surface cavity machining, *Int. Jr. Prod. Res.*, Vol. 30, No. 6, pp.1447-1470

Park S. C. and Choi B. K. (2000), Boundary extraction algorithm for cutting area detection, accepted for publication in *CAD*.

Pratt, M. J. (1984), Solid modelling and the interface between design and manufacture, *IEEE CG&A*, pp.52-59

Shin H., Trecapelli R.M., Shudra Jr. A., Chung Y.C. (1998), A generative approach for die pattern machining, *Proc. of An International IFIP TC5/WG53 Conference*, Machining Impossible Shapes, Chrysler Technology Center, Auburn Hills, Michigan, USA, November 9-11 1998, 24-31

Storr, A., Handel, D., Itterheim, C., Rommel, B. and Strohle, H. (1995), Manufacturing Objects as Uniform NC-Interface for Machining and Measuring, *Proc. of the IFIP WG5.3 PROLAMAT Conference*, Berlin, Germany, pp.356-368

Xu, X. and Hinduja, S. (1998), Recognition of rough machining features in 21/2D components, *CAD*, Vol. 30, No. 7, pp.503-516

Yamazaki, K *et al.* (1995), Autonomous process planning with real-time machining for productive sculptured surface manufacturing based on automatic recognition of geometric features, *Annals of CIRP*, Vol. 44 No. 1, pp.439-444

Yoo Y. S. and Choi, B. K. (1992), CAPP for Die Cavity Machining, *Proc. of the IFIP WG5.3 PROLAMAT Conference*, Tokyo, Japan, pp.437-447

Tolerance Modeling Using the Feature Based Product Model for Prediction of Used Parts Deterioration

M. MATSUDA[1], H. SUZUKI[2] and F. KIMURA[3]
[1] *Department of Information and Computer Sciences, Kanagawa Institute of Technology, 1030 Shimo-ogino, Atsugi-shi, Kanagawa 243-0292, Japan,*
Tel:+81-46-291-3213, Fax: +81-46-242-8490, e-mail: matsuda@ic.kanagawa-it.ac.jp,
[2] *Arthur D. Little (Japan), Inc.*
[3] *Department of Precision Machinery Engineering, The University of Tokyo*

Abstract: The objective of this paper is the prediction for the deteriorated status of mechanical parts based on the feature-oriented product model for supporting the parts reuse phase. As the first step, the product model with "Geometrical Quality Feature" is proposed for representing the deteriorated status. The geometrical quality feature consists of "form tolerance" and "geometrical-relation constraint". Based on the TTRS (Technologically and Topologically Related Surface) method, the form tolerance is described for representing deteriorated deformation of each geometric element. A geometrical-relation constraint is defined corresponding to the type of surfaces and its geometrical relation. The relationship between the form tolerances and the geometrical-relation constraint is also described in each movable element for interpreting the change of form tolerance into that of the geometrical-relation constraint. In the prediction process of parts deterioration using the geometrical quality feature, the effect of the deteriorated deformation on the functional behaviour is calculated using the cause list and behaviour simulation. This calculated status is assigned to each geometric element through form tolerance and it refers to the prediction of the deteriorated status comparing to the usage history. Through this process, the reuse of mechanical parts as a material for the machining process can be supported.

Key words: Life Cycle Engineering, Product Model, Tolerance Modelling, Prediction of Deterioration, Reuse of Parts

1. INTRODUCTION

In order to solve such problems as the shortage of energy and natural resources and environmental pollution, a paradigm change of the present manufacturing system is required. One of the most effective solutions for these problems is to take the whole product life cycle into consideration at the product design phase and realize a closed-loop product life cycle. The product life cycle considered here includes an inverse flow from products to materials or parts. Products are decomposed into several parts after the use phase. Some of these are re-used at the production stage and during the maintenance/up-grade process. In order to reuse parts, their quality has to be predicted geometrically. In order to assist in the quality evaluation of used parts before reuse, a prediction of the status of deterioration with respect to geometrical information like its dimension and tolerance should also be made. In order to support the product life cycle in the view of product information, various product models have been proposed. However, they are mainly focused on the product design and manufacturing phase and its linkage. If the same product model could be used for the other life cycle phases like the prediction of deterioration before reuse, it would lead to a more efficient information exchange throughout the whole product life cycle.

The final objective of this paper is to support the reuse phase by constructing the additional feature of representing the deteriorated status in the feature-oriented product model and by predicting the geometrical quality of each part using this model (Suzuki 2000). In the following sections, the relationship between product life cycle and product model is considered and a *geometrical quality feature* is newly defined in order to represent the deteriorated status with the product model. The geometrical quality feature consists of form tolerances and geometrical-relation constraints. Furthermore the prediction system for parts deterioration based on the product model with geometrical quality feature is proposed. Finally, an example of this method is introduced to verify the feasibility of this method.

2. PRODUCT MODEL FOR PARTS REUSE

2.1 Product life cycle and product model

Due to the rapid development of information technology, product models have been proposed to integrate the various aspects of information about mechanical products (Krause 1993). With the aid of CAD tools, a geometry-oriented product model is created during the product design phase. From the view of each life cycle phase, each phase has a corresponding different viewpoint and acquires necessary information from the same geometry-oriented model. On the other hand, the life cycle of a certain product includes various phases such as planning, manufacturing, use and so on. *Figure 1* shows this relationship between the life cycle flow for each product and the life cycle process, especially focused on the manufacturing phase.

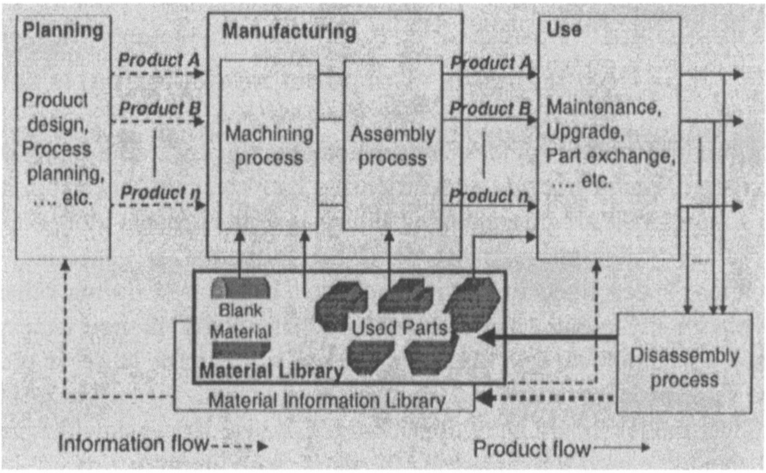

Figure 1. Product Life Cycle with part's reuse flow

In this paper, mechanical products are considered that are produced with large varieties of types and very small production volumes. In this type of production, it is important that used parts from a product can be accepted as materials for another product in the manufacturing phase. Therefore, the manufacturing phase should be focused as well as the reuse phase itself. In the manufacturing phase considering parts reuse, several types of parts from different products have to be processed in the same manufacturing process. The manufacturing process can be divided into machining process and assembly process. If only blank materials are used in the manufacturing process, only one machining process plan can be designed. However, if used parts from different products are also treated as candidates of machining in a material library, several processes should be considered. Because the deteriorated status of each used part can be different, one part may need to be refurbished while others may be able to be used as assembly parts without refurbishment. Also some parts may be used as repair parts in the maintenance process. Here, refurbishment means the additional machining required for remaking the parts into the parts for another product. Many of the mechanical products, which are considered in this paper, have already been intended to be used repeatedly by refurbishment if necessary. For this mixture of product life cycle lines, more complicated processes are required in the manufacturing phase (Matsuda 1999). From the viewpoint of energy consumption, reusing parts as materials for machining tends to consume less energy than in the case of producing from blank material. And during the machining phase, parts reuse makes it possible to decrease the amount of machining, which results in the saving of manufacturing resources. Therefore the parts reuse phase should be involved in the product life cycle of the mechanical products.

2.2 Requirements of product model for parts reuse

The product model provides information about the overall product shape, product size, finishing accuracy, relationship between parts and so on. In this paper, solid models are considered as the basis of the product model. During the product design phase, most of this information is generated. This information is described in the product model and used during other life cycle phases. However, the present product model is not necessarily amenable to support inverse life cycle phases such as parts reuse. In order to use the same product model in the reuse phase as well, the requirements for the product model should be re-considered. *Figure 2* shows how the product model should be used for supporting the parts reuse phase.

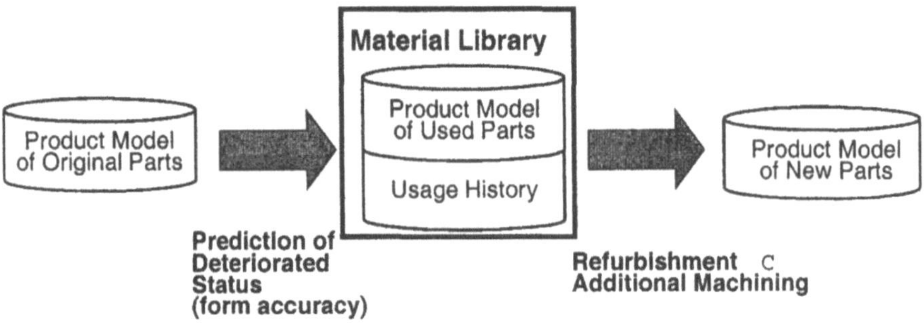

Figure 2. Parts reuse process and product model

The deteriorated condition of each used part is predicted as a change in accuracy of form based on the usage history. An altered product model based on the prediction of deterioration is added to a material library as a candidate for machining. If it matches requirements for another part, refurbishing process data for the selected candidate parts are generated based on the selected product model. Then this product model is again circulated as a new product model in the next life cycle.

For supporting the parts reuse phase, the product model has to be able to be used in the quality management of used parts before its reuse. For this quality management, the product model has to fulfil the following three requirements. The first requirement is the ability to apply the prediction process of geometrical deteriorated status using the calculated method such as a behaviour simulation. As considered, a geometry-oriented product model is mainly applied to predict how a certain geometrical property, such as accuracy of form may have an effect on the product behaviour. In the product model, these geometrical accuracies are represented as the relationship between geometric elements such as surfaces. Therefore, if these properties are involved in the geometrical product model, they should be treated as parameters of the input model in the calculation process. The second requirement is the ability to add the result of the prediction into the original product model. In the present product model, it is hardly supposed that the information about geometrical quality should be added for the inverse process. In order to add the result of the prediction process, another attribute should be added to the normal product model. The third requirement is the ability to be used for production planning including refurbishment. Since such machining is done for each part, these geometrical accuracies have to be assigned to each geometric element from the refurbishment viewpoint.

3. GEOMETRICAL QUALITY FEATURE

From the discussion of the requirements of the product model for employment in the parts reuse phase, it was found that one of the most critical requirements is how to represent the geometrical accuracy in the product model. At present, tolerance concepts have important roles as an indicator of geometrical accuracy during the design, manufacturing and inspection phase of the new parts. Any dimension and shape of machined parts has some geometrical variations from nominal value and shape. While product functions have to be satisfied using parts with geometrical variations. Tolerance concepts have been introduced for showing the amount of tolerant variations from both the functional requirements and the machining cost. (Bjorke 1989) For this role, each tolerance is assigned to each independent part and the designer then sets a certain value and transmits the necessary machining accuracy from the design phase to the manufacturing phase. The present tolerance concept can be divided into dimensional tolerance and geometrical tolerance. The standard dimensional tolerance concept is widely used in the industry and plays an important role for the interchangeability of mechanical parts. However this concept leaves some ambiguities about the shape. On the other hand, geometrical tolerance concepts can give the strict definition of each geometrical form, position and orientation. The present tolerance definitions have not been shown to be general indicators of the geometrical quality at the inverse phase such as parts reuse. In this paper, this tolerance definition should be extended to include geometrical quality through the whole product life cycle. It means that a minute change of geometry due to such deterioration as wear and abrasion should be represented by this extended geometrical tolerance description. For these reasons, a *geometrical quality feature* is newly defined based on the extensive tolerance descriptions and is added to the product model. *Figure 3* shows the structure of the product model with the geometrical quality feature.

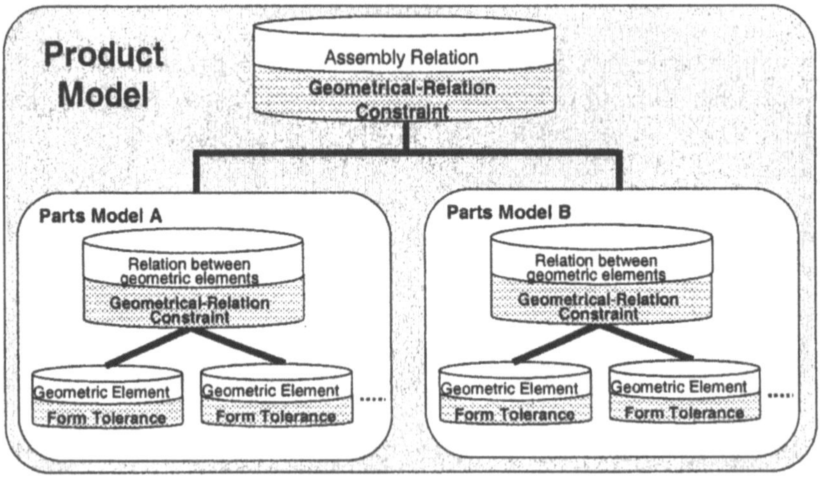

Figure 3. Feature based product model with geometrical quality feature

The geometrical quality feature consists of *form tolerance* descriptions and *geometrical-relation constraint* descriptions. Each form tolerance is described corresponding to a particular geometric element of a part's solid model. On the other hand, geometrical-relation constraints are described with positional and/or orientational relation that represents the connecting relationships between geometric elements. Geometrical-relation constraints are defined based on the present relative tolerance definition. The difference between the normal relative tolerance and this geometrical-relation constraint is that normal relative tolerance is defined between elements which are involved in one part, while geometrical-relation constraints can also be defined between different parts.

Figure 4 shows the role of geometrical quality feature from the product information process. In the mechanical product, many of product functions are realized through the mechanical behaviours. These behaviours are made up of the mechanical capability of each part as well as the part's role in the overall structure. Therefore the geometrical accuracy given by the geometrical variations of each part and its relation can have an effect on the behaviour. The main role of the geometrical quality feature is to relate the mechanical behaviours and the geometrical accuracies.

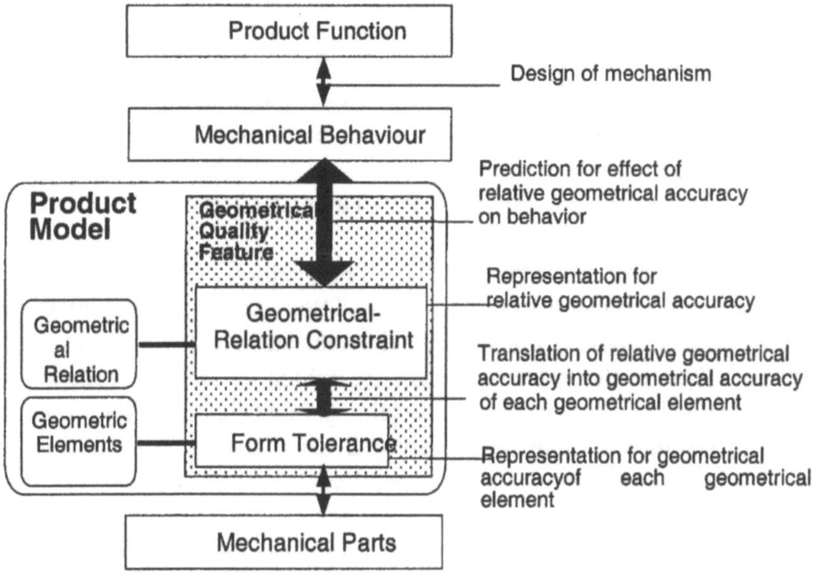

Figure 4. Relationship between product function and its geometry using geometrical quality feature

There are three advantages to be derived from the geometrical quality feature. The first advantage is that by changing the value of form tolerance and geometrical-relation constraint due to deterioration, the geometrical quality feature can be used as a parameter for a geometrical accuracy. The second advantage of this feature is to define the relationship between form tolerance and the geometrical-relation constraint. The third is that both form tolerance and geometrical-relation constraints are based on the definition of the present geometric tolerance, so that the exiting tolerance methods such as tolerance specification and tolerance analysis can be applied for specifying this feature and prediction process.

4. DESCRIPTION OF GEOMETRICAL QUALITY FEATURE

4.1 Description of form tolerance and geometrical-relation constraint

In order to define form tolerance and the geometrical-relation constraint, TTRS (Technologically and Topologically Related Surfaces) (Salomons 1995) is introduced in this paper. In the TTRS method, any surface on the solid model can be divided into seven types based on the displacement that leaves a surface invariance; spherical, planar, cylindrical, helical, surfaces of revolution, prismatic and any surfaces. Table 3 shows these elementary surfaces of TTRS.

One of the biggest advantages of the TTRS method is that it is possible to define new TTRS repeatedly between two TTRS according to the invariance. So TTRS can be applied not only to the real surface involved in a solid model but also to the relative geometry, which is called *Pseudo_TTRS* (Clement 1995). This means a certain geometrical-relation constraint between different parts can be represented using the same definition of a standard relative tolerance such as parallelism, coaxiality and so on. Based on this repetitive TTRS definition, a proper type of geometrical-relation constraint can be inferred. Specific form tolerances can be defined for each elementary surface of TTRS in order to describe its complete shape. If the shape of parts has been changed by such deterioration as abrasion, this change is interpreted as the change of the value of the form tolerance. Each elementary surface of TTRS can be also applied to specific relative tolerance types that can be defined in the TTRS relation involved. In this paper, this is defined as a geometrical-relation constraint based on the relation between the involved TTRS. Table 1 also shows definable geometrical-relation constraints and form tolerances for each elementary surface.

4.2 Description of movable element

To represent the geometrical deteriorated status using the geometrical quality feature, the very part, which had caused the deformation by deterioration, should be paid attention to. As relative motions between contacting geometric elements cause many of the geometrical deteriorations such as wear in a mechanical product, in this paper it is assumed that the deformation caused by deterioration occurs only to movable parts. Therefore, it is important how movable elements are modelled using the geometrical quality feature. In the mechanical product, there are some typical elements for movement. Table 2 shows some examples of these typical movable elements (Suzuki 2000).

Table 1: Elementary surface of TTRS and its definable tolerance

Elementary Surface of TTRS	Definable Form Tolerance	Definable Relative Tolerance (Geometrical-relation constraint)
Sphere	Circularity	Coaxiality
Plane	Flatness, Straightness	Parallelism, Perpendicularity, Position
Cylindrical	Cylindricity, Circularity, Straightness (Axis)	Coaxiality, Parallelism, Perpendicularity, Position
Helical	(None)	Coaxiality, Parallelism, Perpendicularity, Position
Of revolution	Profile of surface, Circularity, Straightness (Axis), Profile of line	Coaxiality, Parallelism, Perpendicularity, Position
Prismatic	Profile of surface, Straightness (Axis), Profile of line	Coaxiality, Parallelism, Perpendicularity, Position
General	Profile of surface, Profile of line	(None)

Table 2: Example of common movable elements

Example of Movable Elements	Combination of Surface Type	Type of Geometrical-relation Constraint	Type of Movement	Applied Form Tolerance
Feed Elements 	Cylinder – Plane	Parallelism	Translation [Plane] + Rotation [Cylinder]	Flatness, Cylindricity
Crank Element 	Cylinder – Cylinder	Coaxiality	Translation (along axis)	Cylindricity
Slider Element 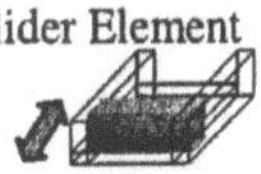	Prismatic – Prismatic	Coaxiality, Parallelism	Translation (along axis)	Profile of Surface
Hinge Element 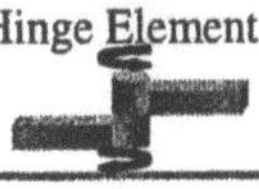	Cylinder – Cylinder Plane – Plane	Coaxiality, Parallelism	Rotation (along axis)	Circularity, Flatness

Each movable element consists of two geometric elements. Each surface of the elements is applied to the appropriate elementary surfaces of TTRS and then a proper form tolerance is defined for each surface based on it. As these contact surfaces also have some geometrical relations between themselves, a geometrical-relation constraint can be also defined between elements using the TTRS method. This geometrical-relation constraint has the role to indicate the required accuracy in that movement. These movements occur along the direction in which TTRS has a degree of freedom.

4.3 Relationship between form tolerance and geometrical-relation constraint

After predicting the deteriorated status using a geometrical-relation constraint, the result has to be assigned to each geometric element. For this assignment, the relationship between a form tolerance and a geometrical-relation constraint should be defined in a movable element. *Figure 5* shows an example of this relationship in the case of a hole and a pin.

Relation Table

Relation:Coaxial (Concentrical)

	[S]	[Pl]	[C]	[H]	[R]	[Pr]	[G]
Sphere	○	×	○	○	○	○	×
Plane		×	×	×	×	×	×
Cylindrical			○	○	○	○	×
Helical				○	○	○	×
Of revolution					○	○	×
Prismatic						○	×
General							×

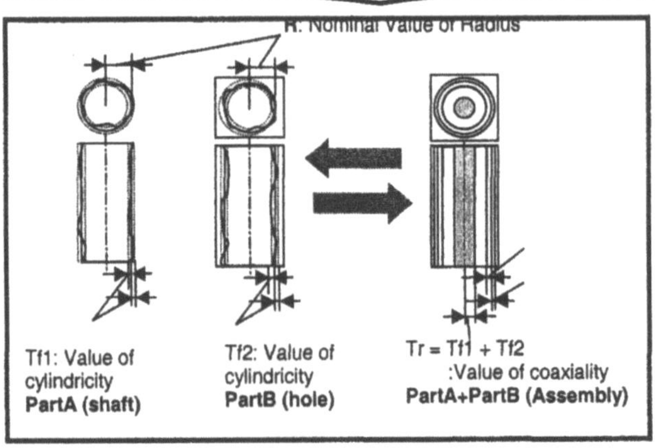

Figure 5. Relationship between form tolerance and geometrical-relation tolerance

When the geometrical-relation constraint based on the type of form tolerances and its relation is defined, the relational table as shown on the left hand side of *Figure 5* is referred to. In this table also describes this relationship between the form tolerances and the geometrical-relation constraint. As both surfaces are cylinders in this example, both part A (shaft) and part B (hole) have the same radius R as a nominal value and these deviations can be restricted by cylindricity. (The values of cylindricities are Tf1 and Tf2 respectively.) Then the shaft and hole are assembled. In the assembly, it is considered that both parts have the shape that provides maximum clearance between them. Under this assumption, the shaft can move freely within the hole due to the clearance. This can be interpreted as the centre of the shaft can move within the cylindrical area whose radius is Tf1+Tf2. This means that the value of coaxiality between the shaft and the hole is Tf1+Tf2. This is the relationship between two cylindericities (form tolerance) and a coaxiality (geometrical-relation constraint). Using the above-mentioned relation, the minute deformation of each surface caused by deterioration can be related to the geometrical-relation constraint through the change of form tolerance value.

4.4 Representation of geometrical quality feature

Based on the method that was explained in the previous three sections, the geometrical quality feature is implemented in the product model. *Figure 6* shows this description and the relation.

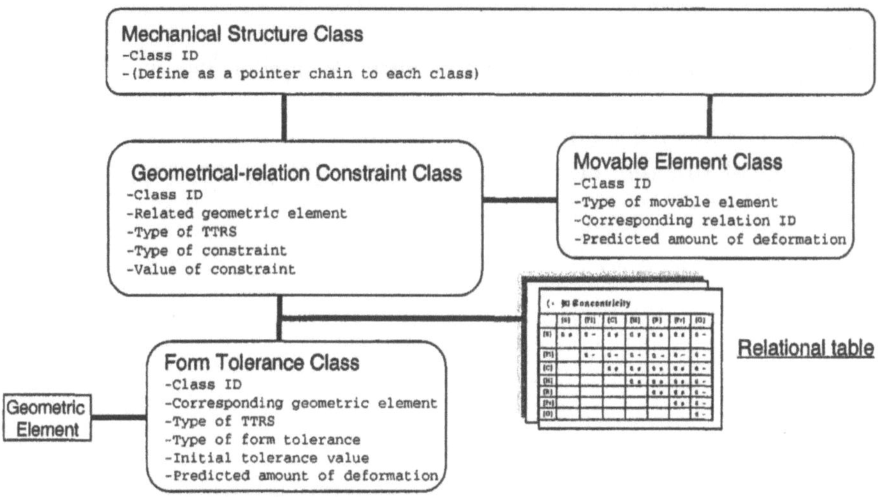

Figure 6. Description of geometrical quality feature

In the implementation using C++, each element is defined independently as a class object. At first, form tolerances are defined to each geometric element. The type of form tolerance is determined according to the surface type of the geometric element. This form tolerance class has an attribute for adding the amount of deteriorated deformation based on the result of the prediction process. Secondly, a geometrical-relation constraint is defined to geometrical relation between geometric elements. Using the surface type involved the relation and the geometrical relation such as position and orientation, the type of geometrical-relation constraint is specified from the relational table based on the TTRS. If the geometrical-relation constraint is defined between parts and any relative motion occurs in the relation, a movable element has also to be defined to the geometrical-relation constraint. As commonly used movable element classes have already been constructed, the appropriate movable element is applied to the objective movable relation. This class also has an attribute for representing the deteriorated status based on the prediction result. Finally, the whole connection between each class in the product is described as a mechanical structure class. In this class, the connective relation is represented in the form of a pointer chain between each class.

5. PREDICTION SYSTEM OF DETERIORATION STATUS

5.1 Calculation process using cause list

Using the product model with the geometrical quality feature, the effect of the geometrical accuracy on the functional behaviour of the product is calculated. For this calculation process the *cause list* is generated based on the mechanical structure class. *Figure 7* shows this process and the cause list.

A mechanical structure class represents the relationship between each element of the geometrical quality feature. In the existing tolerance specification method and tolerance analysis method, a tolerance chain is defined in order to consider the interrelation between each tolerance (Salomons 1995). Therefore this mechanical structure class can be regarded to be the same as a tolerance chain, while, in this paper, functional requirements for satisfying the product function are defined as the geometrical accuracies of assembly relations, behaviour tracks and so on. Corresponding to each functional requirement, a cause list is generated. In this list, the possible sets of the movable elements in which deteriorations occur are given corresponding to a functional requirement to be achieved. In order to decide whether the functional requirement is achieved or not, it is assumed that the threshold value has been given by the designer in advance.

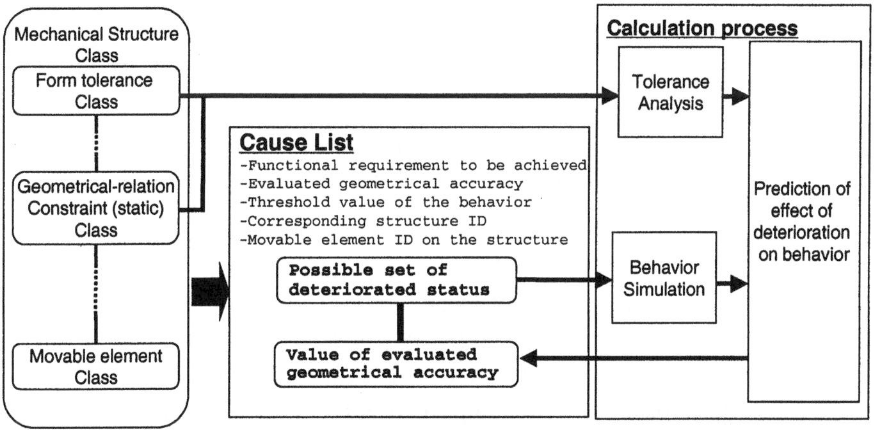

Figure 7. Calculation process using cause list

In the calculation process, each set of deteriorated status given by the cause list is tried with the behaviour simulation. In the behaviour simulation, the value of the corresponding geometrical-relation constraint in the movable element is interpreted as a dimension of the clearance between parts. On the other hand, the other form tolerances and geometrical-relation constraints, which are not related to the deterioration in the use phase, may also have effects on the behaviour due to the initial tolerances. These effects can be calculated using the existing tolerance analysis method in a statistical way. By superimposing the result of behaviour simulation and tolerance analysis, the effect of that set of deteriorated status on the behaviour can be calculated as the value of the corresponding evaluated geometrical accuracy. These results are added to the cause list based on the calculated result.

5.2 Prediction process of deteriorated status

After the calculation process using the cause list mentioned in the previous section, a prediction process for the deteriorated status of parts is proposed as shown in *Figure 8.*

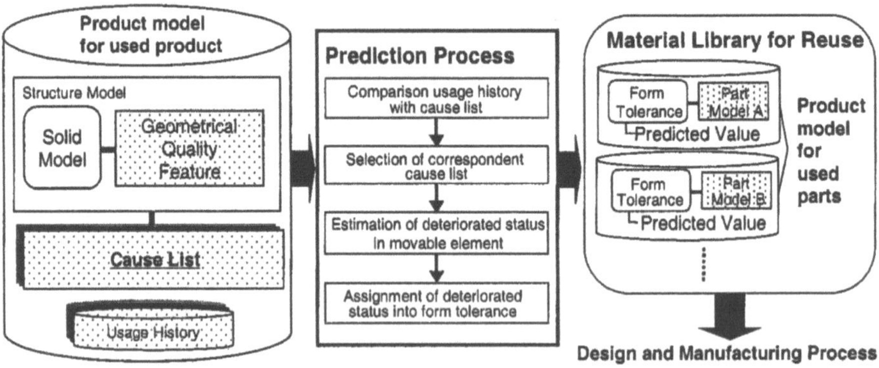

Figure 8. Product Model Based Prediction process for parts deterioration based on the product model

For the prediction process, geometrical quality feature and cause list have been added to the product model for used parts during the product design phase and the use phase. Usage history is also added to the product model during the use phase. This history mainly describes the functional status of the product corresponding to the functional requirement on the cause list.

For the prediction, as the first step the usage history is compared with the threshold value of each functional requirement on the cause list. If a functional requirement to be achieved is found to be satisfied from the usage history, the cause list corresponding to the functional requirement can be selected for predicting the deteriorated status of related movable elements. Based on this, the deteriorated status for each movable element is estimated. Then in each movable element, the deformation is assigned to the form tolerance of each geometric element involved in the movable element. The relationship between geometrical-relation constraint and form tolerance is used for this assignment. Finally, to each used parts model is added the predicted value under the deteriorated status. The used parts model with deteriorated status is added to the material library for the machining phase together with blank material. If it matches the new specification for machining, it can be considered as a candidate for material to be selected. Therefore, each used part can be reused effectively.

6. EXAMPLE

In this chapter, the proposed prediction method is applied to a spindle mechanism. *Figure 9* shows this mechanism and its functional requirements and movable elements. In this model, there are two movable elements; left bearing and right bearing. Two functional requirements should be satisfied during usage; the positional error of movement at the edge of the gear both in the axial and radial direction.

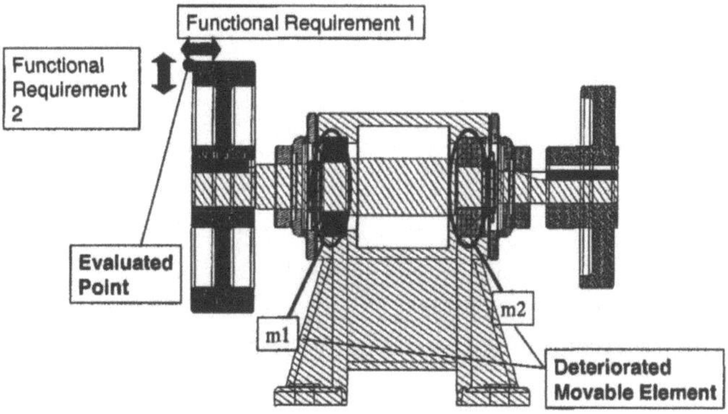

Figure 9. Spindle mechanism

Based on this solid model, both form tolerances and geometrical-relation constraints are added. *Figure 10* shows the description of the geometrical quality feature relation of this model. If a movement occurs in the relation, an appropriate movable element is also defined. In this example, a hinge element class is defined between the outer race and inner race of both bearings. By repeatedly generating geometrical quality features for all parts, the mechanical structure class can be defined.

A cause list is then generated based on the mechanical structure and a functional requirement. Then each possible set is checked in the calculation process. In this process, behaviour simulation is executed in order to estimate the evaluated geometrical accuracy for each functional requirement. In the prediction process, if this function requirement is satisfied, the corresponding cause list is selected for predicting the worst deteriorated status. In this case, it is represented as the dimension of clearance in this case, at each bearing element as shown in *Figure 11.*

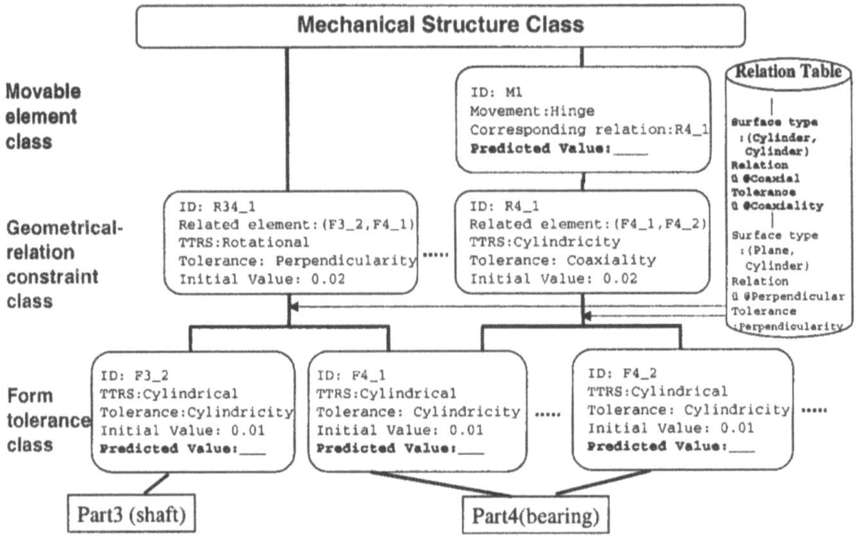

Figure 10. Description of geometrical quality feature of the spindle model

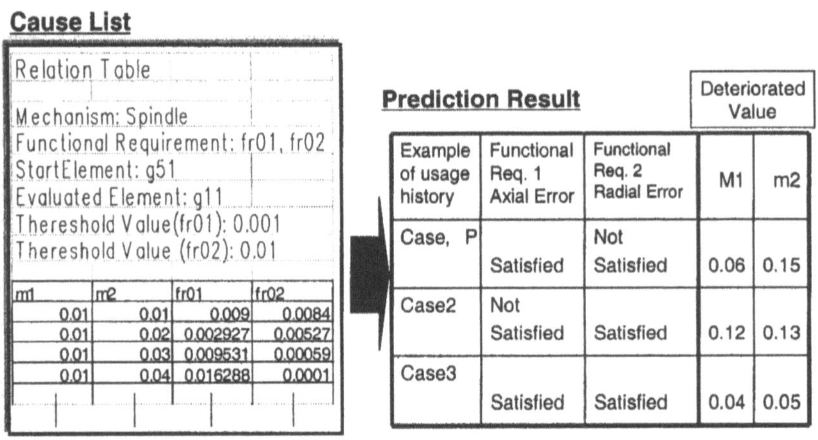

Figure 11. Prediction of deteriorated status of movable element

7. CONCLUSIONS AND DISCUSSION

In this paper, a reuse phase for the mechanical parts could be supported successfully by defining a geometrical quality feature in the feature-oriented product model and by predicting the deteriorated status of parts using this feature. The geometrical quality feature was described by semantically extending the concept of the present geometric tolerance. It consisted of form tolerance description, geometrical-relation constraint and the relationship between them. It can be said that this structure has several advantages. The first advantage is that the geometrical quality feature is defined based on the definition of the present geometric tolerance. Therefore, tolerance specification methods, such as the TTRS method, can be applied to that of geometrical quality features. The second advantage is that both of the parts deformation due to deterioration and the change of the assembly relation can be represented using form tolerance descriptions and geometrical-relation constraints respectively. Furthermore by defining the relationship between form tolerance description and geometrical-relation constraint, functional requirements about geometrical accuracy of assembly relations could be interpreted into the form accuracy of physical parts for the refurbishment process.

Using the product model with the geometrical quality feature, a prediction process of deteriorated status could also be proposed. In executing a behaviour simulation for the prediction, geometrical-relation constraints could be treated as geometrical parameters. As a result, the deteriorated status in the movable element be calculated and reflected in terms of the form tolerance. This information is added to the product model and thus helps in the prediction process by comparing the usage history. Finally, the example process for the prediction of deteriorated status could be shown. This prediction process will lead to the creation of an efficient material library including used parts. The material library will make it possible to construct the process planning system including the refurbishment process for used parts. This means that the proposed method permits the construction of feature-oriented product model that could support the realization of a closed product life cycle.

ACKNOWLEDGMENT

The authors are grateful to Dr. Udo Graefe of the National Research Council of Canada for his helpful assistance with the writing of this paper in English. The authors also thank the Life Cycle Group member in the Kimura Laboratory of the University of Tokyo for a fruitful discussion.

REFERENCES

Clement, A., Rivière, A. and Serre, P. (1996). A Declarative Information Model for Functional Requirements, In Kimura, F. (ed.): *Computer-aided Tolerancing, Proceedings of the 4th CIRP Design Seminar,* pp.3-16.

Krause, F. -L., Kimura, F., Kjellberg, T., and Lu, S. C. –Y. (1993). Product Modelling, In Annals of the CIRP Vol. 42/2/1993, pp. 695-700.

Matsuda, M., and Kimura, F. (1999). A Model Integration Framework for Life Cycle Engineering from the Machining Process Viewpoint, In CIRP 6th International Seminar on Life Cycle Engineering, pp. 332-341.

Oyvind Bjorke (1989). *Computer-aided Tolerancing*, ASME Press.

Salomons, O. (1995). Computer Support In The Design Of Mechanical Products, PhD Thesis of the University of Twente.

Suzuki, H., Matsuda, M. and Kimura, F. (2000). Prediction of Used Parts Deterioration Based on A Tolerance Model, In CIRP 33rd International Seminar on Manufacturing Systems, pp. 149-154.

Product Recovery-Cycles Design
Extension of Useful Life

F. GIUDICE, G. LA ROSA and A. RISITANO
DIIM Department of Industrial and Mechanical Engineering, University of Catania, Viale A. Doria 6, 95125 Catania, Italy - e-mail: fgiudice@diim.unict.it

Abstract: This paper reports the development of a tool to aid designers in making the best choices in order to plan recovery cycles for a product at the end of its working life. Having outlined an effective design methodology for this purpose, a calculation model was developed to support the methodology. The model was based on the quantification of determining factors, allowing the definition of the reusable parts of the architecture and the evaluation of the values assumed by an indicator which translates the environmental effects of recovery cycles in terms of extension of the product's useful life. A trial conducted on a commonly used industrial unit, a heat exchanger, demonstrated the value of the design tool in comparing different architectures of a product and their optimisation, confirming its effectiveness as calculation model for the environmental quality-oriented product design.

Key words: Life-Cycle Design, Design for Environment, Recovery Cycles, Intensity of Use, Useful Life

1. INTRODUCTION

The continuous search for ever greater competitive advantage in industry has led to the development of products which offer not only high quality standards throughout their working life, but also a wide range of properties characterising their performance in different phases of their whole life cycle.

These production requirements have influenced the evolution of design methodologies which reflect how initial design choices can greatly affect a product's performance over its entire life cycle, in terms of functionality, ease of production and servicing, and environmental impact (Kusiak 1992, Ishii 1995, Huang 1996).

In the context of these new methodological approaches, known as Life-Cycle Design or Design for X (DFX), Design for Environment (Navin-Chandra 1991, Billatos 1997, Graedel 1998) is directed at providing products with an elevated level of eco-compatibility over their entire life cycle. Its principle objective, that is to optimise the distribution of the flows of resources over the product's whole life cycle, can be followed by many strategies. Of particular importance are those concerning the recovery of components and materials at the end of the product's working life (disassembly, selection, inspection, direct reuse of components, recycling material in primary or external production cycles), in that they are directed at closing the resource flow cycles through recycling (*Figure 1*).

These strategies translate into design approaches which target disassembly, recovery and recycling, known as Design for Disassembly and Recycling (Burke 1992, Dowie 1995, Ishii 1999).

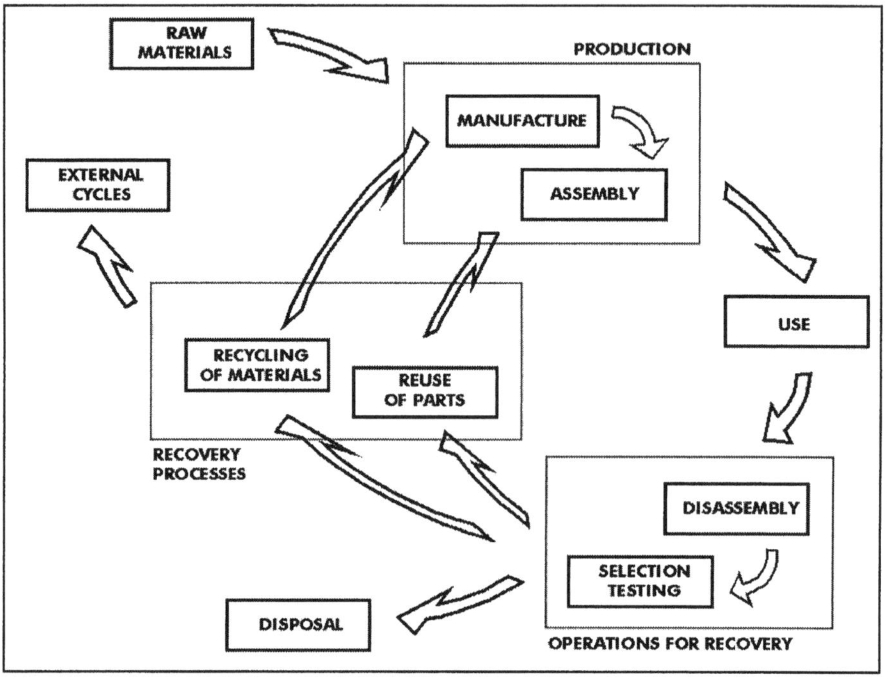

Figure 1. Life cycle and resources recovery strategies

2. DEFINITION OF THE PROBLEM AND AIMS

Recovery operations at the end of a product's life can be divided into the following main phases (*Figure 1*):

- disassembly of parts;
- selection and inspection of their condition;
- direct reuse of complete components;
- recycling materials in the primary production cycle or in external cycles.

Thus the suitability and effectiveness of these operations are strictly dependent on several determining factors which characterise the architecture of the constructional unit:

- the detachability of components (depending on the layout of the unit and on the geometry and the junctions between parts);
- the property of component durability, which conditions the process of declining performance over time, and consequently determines their suitability for reuse;
- the appropriateness and suitability of recycling material from parts are not reusable.

On the basis of the experience of manufacturers operating in the industrial plant sector, for some product typologies, it seems that generally resources are used at a low intensity in that some parts of a constructional unit were still efficient at the end of its conventional life span and, therefore, potentially could guarantee further durability, but were nevertheless thrown out with the entire product. This evidenced interesting margins in which operations of component reuse could be applied. In the light of these considerations, the research was orientated towards the development of a tool able to translate some determining design choices into opportunities for component recovery by means of a mathematical model which simulates possible recovery cycles at the end of the product's working life and quantifies the benefit in environmental terms.

The aim, therefore, was to develop a design tool which allows architectures to be compared and optimised, in terms of component separability and the distribution of their useful life in the unit, for an effective recycling of resources.

3. DESIGN METHODOLOGY

The problem under examination and the stated objective suggested the development of a design methodology which can be schematised as in *Figure 2*.

The first phase, defining the overall architecture of the product using conventional design tools, consists of identifying the principle components of the system, their spatial and functional arrangement, jointing systems, and the overall geometry and materials of each component.

This is followed by a phase in which the conventional architecture so defined was analysed in relation to the recovery strategies. This analysis seeks to interpret the architecture in terms of component modules, highlighting the property of durability of performance in time and the jointing restraints (reversible/irreversible) which condition the possibility of separating a component from the rest of the unit. With regard to this last point, a qualitative analysis based on simple observation of the system's architecture can be complemented by an analytical approach to the problem of evaluating the separability of each component, or module, using appropriate models for the analysis of disassembly plans. This allows the design method to be further broken down, integrating it with the tools of Design and Planning for Disassembly, which are directed at optimising the disassembly procedure by reducing the associated costs, operating on the modularity and the choice of jointing systems in the unit (Navin-Chandra 1994, Lambert 1997, Johnson 1998, Tang 2000).

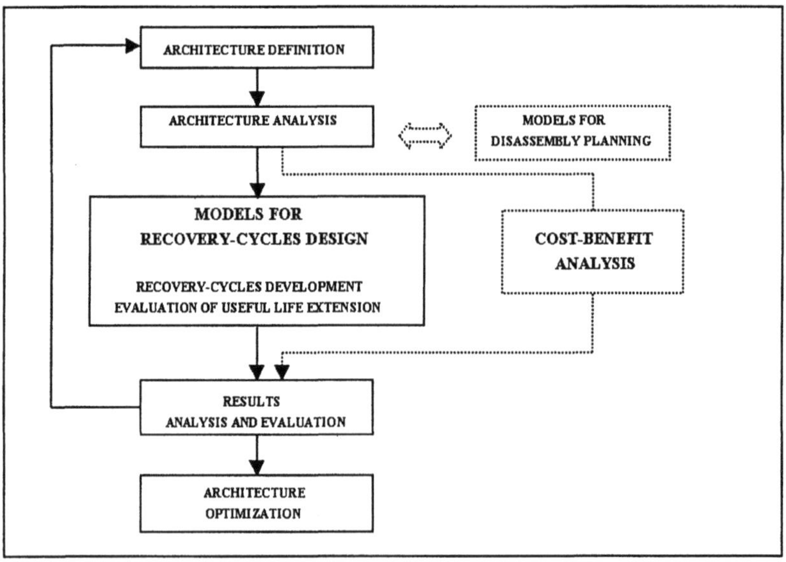

Figure 2. Design methodology

The data obtained in the architecture analysis is then elaborated using the calculation model proposed here which develops and analyses the possible recovery cycles, and evaluates their environmental effectiveness by means of an indicator which quantifies the intensity of resource use in terms of extension of the product's useful life. This approach, which allow the optimisation of the product's environmental performance, could be complemented by an analysis which evaluates the not insignificant life-cycle costs related to the recovery strategies. The analytical formulation of the problem allows the developed tool to be integrated with others for the cost-benefit analysis of recovery processes (see *Figure* 2), such as those already proposed by the authors (Giudice 1999a). This specific problem will be subject of further works.

The final phase consists of evaluating the results obtained with applied calculation models, which can provide suggestions for a correct redesign of the architecture to obtain a definition which manages to harmonise the requirements of conventional design (functionality, safety, costs) with environmental demands.

4. CALCULATION MODEL

The calculation model proposed here, developed as an aid to the design of product recovery cycles, has the function of analysing the architecture data, translating this into the suitability of a component for reuse, and calculating the outcome of the possible recovery cycles at the end of the product's ordinary working life, quantifying the advantages in terms of value of resources used in production, and consequent environmental protection.

4.1 Basic procedure of the calculation model

The calculation model, represented in the design methodology (*Figure* 2), is based on a procedure summarised in *Figure 3*.

The data supplied by the analysis of the architecture is initially elaborated to identify the determining factors for component recovery, i.e. the property of each component which determines the possibility and suitability of its recovery and reuse. As we have seen, these factors were identified as the predicted duration of functional efficiency and safety of each component, and as its separability from the rest of the unit. It is then possible to evaluate the reusability of the components, i.e. the possibility of reusing them one or more times, and consequently to plan all the possible recovery cycles. The final phase of the calculation procedure translates the recovery plans into intensity of use of the resources, introducing an indicator of extension of the product's useful life.

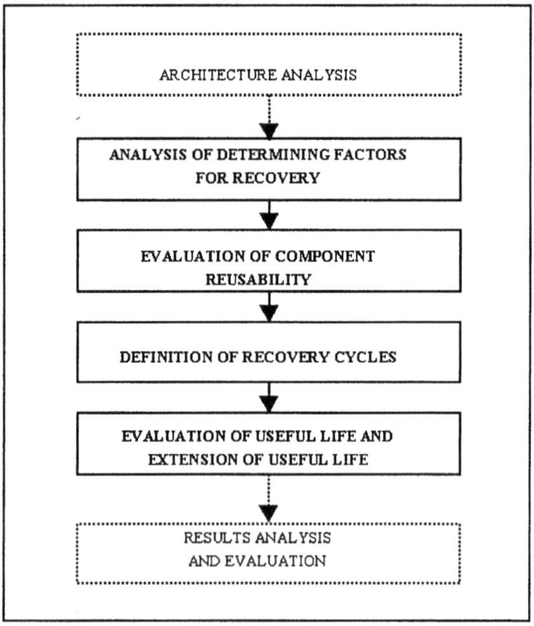

Figure 3. Basic procedure of calculation model

4.2 Determining factors for recovery

The analysis of the overall architecture, conceived according conventional design criteria, allows the identification of n principle components C_i making up the unit, each characterised by weight P_i and material M_i. Having identified the components, which may be grouped in construction modules, two factors determining their suitability for recovery and reuse must be defined: durability and separability.

The durability D_i of the i-th component is the ratio between the predictable duration of component DP_i, and the duration of the product's conventional working life T:

$$D_i = \text{int}\left(\frac{DP_i}{T}\right) \tag{1}$$

The terms DP_i depend on design choices, and are quantifiable on the basis of modern design methods, which make it possible to guarantee the complete efficiency of the component in time, and on the basis of information obtained from the past behaviour of pre-existing components. The duration T is the time predicted for the normal use of the entire product without the substitution of any part.

The separability of the i-th component expresses the possibility of disassembling and separating the component from the unit in order to recover it. This depends not only on the lay-out of the architecture and on the jointing constraints, but also on the relation between the component in question and other components characterised by the same durability, because, in recovery terms, a component is separable if it can be disassembled from the unit, or if it belongs to a group of components characterised by the same durability, and therefore constituting a single recoverable sub-unit, which can be disassembled.

To better define the separability, from (1) it is possible to find the number of potential reuses R^*_i of each i-th component on the basis of its durability:

$$R^*_i = D_i - 1 \tag{2}$$

Consequently, the i-th component potentially takes part in the j-th recovery cycle if the condition $R^*_i \geq j$ is true. From the verification of conditions of this type it is possible to identify the group of components which, for durability, could be included in each potential recovery cycle.

The separability S_{ij} of the i-th component at the j-th recovery cycle is thus defined as follows: S_{ij} assumes a value 1 if it is possible to disassemble the single i-th component or the group of components potentially reusable at the j-th recovery cycle to which the i-th component belongs, otherwise it is equal to 0.

4.3 Determining factors matrices and recovery vectors

Indicating by m* the overall number of potential recovery cycles ($m^* = \max(R^*_i)$), it is then possible to define the m* matrices of the determining factors for each j-th recovery M^{FR}_j, which collects the durability and separability factors of all the components:

$$M_j^{FR} = \begin{bmatrix} D_i & S_{ij} \end{bmatrix}_{i=1,2,\ldots,n} \tag{3}$$

and the corresponding recovery vectors V^R_j

$$V_j^R = \begin{bmatrix} V_{ij}^R \end{bmatrix}_{i=1,2,\ldots,n} \tag{4}$$

which collect the n terms V^R_{ij} defined thus:

$$V_{ij}^{R} = D_i \cdot S_{ij} \tag{5}$$

For the i-th component to be effectively reusable at the j-th recovery cycle, the condition $V^{R}_{ij} \geq 2$ mast be true (because $D_i \geq 2$ and $S_{ij} = 1$ mast be true).

4.4 Effective component reusability

The effective reusability of the i-th component can thus be expressed in two ways.

The relative reusability r_{ij}, which expresses the reusability of the i-th component at the j-th recovery cycle, is defined on the basis of condition $V^{R}_{ij} \geq 2$ as follows:

$$r_{ij} = \begin{cases} 0 & \text{if} \quad V_{ij}^{R} < 2 \\ 1 & \text{if} \quad V_{ij}^{R} \geq 2 \end{cases} \tag{6}$$

and therefore assumes a value 1 if the i-th component is reusable at the j-th recovery cycle, otherwise it is equal to 0.

The absolute reusability R_i, which expresses the total number of recovery cycles which the i-th component can undergo after its first conventional use, is then given by:

$$R_i = \sum_{j=1}^{m} r_{ij} \tag{7}$$

where m is the maximum number of foreseeable recovery cycles for the entire unit, function of m* and of r_{ij} terms. If $R_i = 0$ the i-th component is not reusable, if $R_i \geq 1$ the component can be reused a further R_i times.

4.5 Recovery fractions

The recovery fraction Φ^{P}_{j}, which expresses the fraction of the total weight of the product reusable at the j-th recovery cycle, is expressed by:

$$\Phi_j^P = \frac{\sum_{i=1}^{n} r_{ij} \cdot P_i}{\sum_{i=1}^{n} P_i} \tag{8}$$

These terms define the quantity of material in play in each recovery cycle and therefore already give a quantitative indication of the efficacy and the benefits of the recovery operations. To better characterise these in terms of environmental impact, however, it is appropriate to express them according to the environmental impact associated with the production of each component. This is possible using the environmental indicators available in the literature. Applying specific methodologies, these indicators are used to quantify the impact involved in the production of the more common materials and in standard manufacturing processes (Goedkoop 1995). Indicating the environmental impact of i-th component production by α_i, this can be expressed as:

$$\alpha_i = EI_i^{mat} \cdot P_i + \sum_{s=1}^{t_i} EI_i^{pr_s} \cdot p_i^{pr_s} \tag{9}$$

where

- EI^{mat}_i is the unitary eco-indicator for the material making up the i-th component;
- P_i is the weight of the i-th component;
- $EI_i^{pr_s}$ is the eco-indicator which can be associated with the s-th manufacturing process of the i-th component's production cycle, per unit of principle process parameter;
- $p_i^{pr_s}$ is the principle process parameter of the s-th manufacturing process of the i-th component's production cycle (e.g. in the case of a cutting process, $p_i^{pr_s}$ is equal to the length of the cut).

The recovery fraction Φ^{α}_j, which expresses the fraction of the product weight reusable at the j-th recovery cycle in terms of environmental impact, is expressed by:

$$\Phi_j^{\alpha} = \frac{\sum_{i=1}^{n} r_{ij} \cdot \alpha_i}{\sum_{i=1}^{n} \alpha_i} \tag{10}$$

The terms expressed by (8) or (10) define the recovery cycles in that they are functions of the components which take part in each recovery cycle.

4.6 Extension of useful life

Useful Life, intended as a period of time for which the entire architecture or part of it is used in the same life cycle, can be considered as an indicator of saving resources. Extending the useful life, in fact, corresponds to a better use of the resources involved in the production phase, and to a safeguard against the excessive dumping of waste.

Already formulated by the authors (Giudice 1999b), on the basis of previous suggestions (Navin-Chandra 1991), this indicator provides a global vision of the product's entire life, beyond the limits of a single life cycle, and is introduced and developed here as a complement to the calculation model proposed in this study.

Summarising some concepts already discussed, therefore, if at the end of its conventional use, of duration T, the product is not recovered in any way, nor reused as a whole, then its useful life is equal to time T during which it performs the function related to that single use. If, instead, the whole product is reused, or if it is recovered in part, the useful life is extended, since all or parts of the resources employed continue to perform the anticipated function.

Referring to precedent formulation (Giudice 1999b), indicating the number of whole reuses by nr, and the number of recoveries by m, if at the conclusion of each use subsequent to integral reuses, a recovery fraction Φ_j is recovered and again introduced into the cycle, hypothesising a constant duration of all the uses equal to T, the useful life is then:

$$UL = \left[1 + nr + \sum_{j=1}^{m} \Phi_1 \cdot \Phi_2 \cdot ..... \cdot \Phi_j \right] \cdot T \qquad (11)$$

where Φ_j is the recovery fraction at the j-th recovery cycle. In the real case the number of reusable components coming from the original unit decrease with each successive recovery.

From (11) one obtains the Extension of Useful Life, which quantifies the overall extension of the life of the original components within the same product's life cycle:

$$EUL = \left[nr + \sum_{j=1}^{m} \Phi_1 \cdot \Phi_2 \cdot ..... \cdot \Phi_j \right] \cdot T \qquad (12)$$

From (12) it is simple to define a final indicator which quantifies the extension of useful life after the generic k-th recovery:

$$EUL_k = \left[nr + \sum_{j=1}^{k} \Phi_1 \cdot \Phi_2 \cdot .... \cdot \Phi_j \right] \cdot T \tag{13}$$

According to the type of investigation required, the recovery fractions expressed by (8) or (10) are introduced into (11) , (12) and (13). In the first case it is just necessary to evaluate the weights P_i of each i-th component. In the second case it is necessary to evaluate the terms α_i expressed by (9) (see also 5.3 section).

5. CASE STUDY: ANALYSIS AND OPTIMISATION OF HEAT EXCHANGER ARCHITECTURES

The model developed for the analysis and design of recovery cycles was applied to a specific design case: the analysis and optimisation of the architecture of heat exchangers. The specific product seems particularly interesting in relation to the problem in question in that: it is characterised by construction standards developed according to principles of modular architecture which makes the product highly suitable for the recovery of some components at the end of its working life; the experience of the manufacturers has shown wide margins for the application of recovery operations at the end of the product's life.

5.1 Construction standards of heat exchangers

The construction of heat exchangers is codified and standardised by clear design guidelines. The definition of the architecture is directed by its subdivision into three main construction modules (two heads and the central body), each of which can be chosen from a range of different design alternatives (TEMA 1988). The various possible combinations in the choice of modules gives rise to a wide variety of architectures.

5.2 Operations for recovery at the end of working life

For some typologies of exchangers, and the uses they are destined for, a non-homogeneous deterioration was found in the metallurgic properties of the steel used. Consequently some longer lasting components would be reusable if they were separable from the constructional unit and if they were considered perfectly efficient from the point of view of their mechanical and thermodynamic properties.

An investigation into the potentiality of the component recovery strategies was conducted with particular regard to two typologies of heat exchanger, with the common parameters of overall volume (shell side capacity 3500 litres, overall length 7-8 m), thermal-pressure working conditions (about 20 bar/400° shell side, about 40 bar/300° tube side) and foreseen use in the industrial plant (shell side fluid VPS-BTMS, tube side fluid R-CRUDE). The two architectures differ in that the first has a detachable shell, coded CFU (*Figure 4*), while the second is of the floating head type, coded AES (*Figure 5*). In both cases, after about 6 years use metallographic analysis showed considerable deterioration in the metallurgical properties of the central module shell, good condition of the tubes (potentially reusable after rigorous cleaning) and excellent condition of the remaining components (front and rear end modules). In analogous tests, after a further 6 years use of exchangers regenerated by substituting only the external central module, the reused tubes showed deterioration, while the good condition of the remaining components suggested their suitability for a further and ultimate reuse.

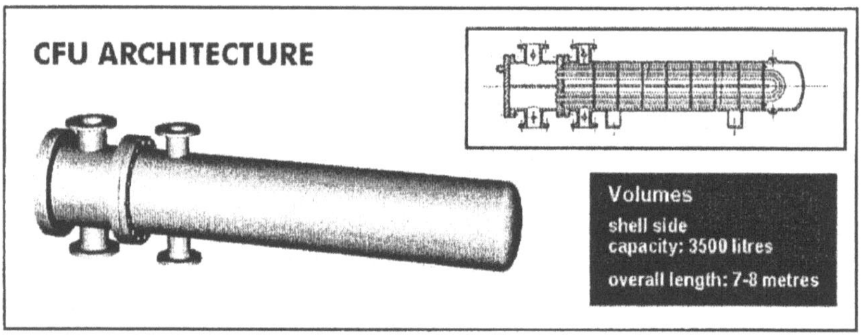

Figure 4. CFU architecture typology

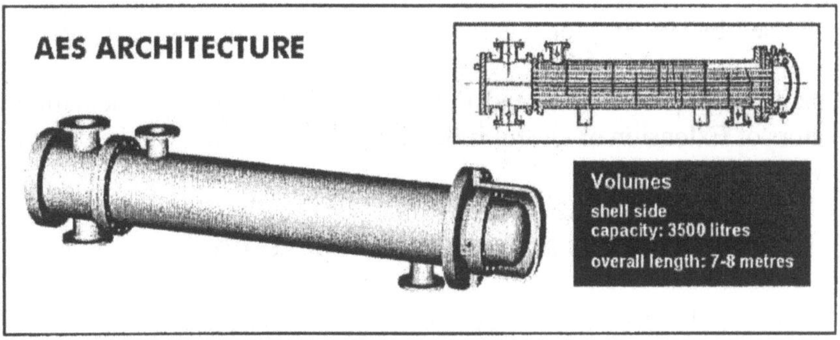

Figure 5. AES architecture typology

The data obtained confirm, therefore, the suitability of applying the recovery cycle analysis developed here in order to evaluate the efficacy of the proposed model in relation to two different design operations:

- comparison between different architecture typologies to evaluate the recovery potential;
- evaluation of design choices aimed at optimising an architecture typology.

5.3 Application of the calculation model

The calculation model was applied following the procedure reported in *Figure 3*.

Analysis of the architectures under examination allowed the identification of the principle components and construction modules. The recovery determining factors were defined making matrices of type (3), and this allowed the definition of two possible recovery cycles and the components or groups of components which can be included on the basis of their properties of durability and separability.

In particular, the predicted duration of each component was that found for the pre-existing components on the basis of the metallographic testing data reported above. The duration of conventional use T was fixed at 6 years, according to the indications of the same data. Component separability was identified on the basis of a simple analysis of the typologies of jointing between the components, according to the distinction reversible/irreversible.

The recovery fractions were calculated in terms of environmental impact, as in (10), because of their greater significance from the environmental viewpoint. The environmental indicators developed according to the Eco-indicator 95 methodology (Goedkoop 1995) were used in the calculation of

α_i expressed by (9), and for this reason the fractions $\Phi^{\alpha}{}_{j}$ were tabulated as $\Phi^{EI}{}_{j}$.

Finally, expressions (12) and (13) were applied for the calculation of indicators of Extension of Useful Life, giving results which will be discussed below.

5.4 Analysis and evaluation of results

As above, the applications developed to evaluate the efficacy of the proposed model, followed two different types of design operation: one of comparison and the other of optimisation. In the first case, the aim was to compare the two architecture typologies CFU and AES (some of the data for the two architectures are reported in Table 1 and 2), to evaluate their different recovery potentialities.

Table 1. Data on the CFU architecture

CFU ARCHITECTURE	QUNATITY	MATERIAL	WEIGHT [kg]	WEIGHT (TOT) [kg]	EI MAT (UNIT) [mPt]	EI MAT [mPt]	EI MAT (TOT) [mPt]
CENTRAL MODULE + REAR END HEAD MODULE							
SHELL	1	FE410.1KW	3953	3953	4,3	16998	16998
SHELL NOZZLE	2	ASTM A105 / A106	44	88	4,1	180	361
SHELL COVER	1	FE410.1KW	341	341	4,3	1466	1466
SHELL FLANGE	1	ASTM A105	574	574	4,1	2353	2353
** SUPPORT SADDLE	2	FE410.1KW	101	202	4,3	434	869
				5158			22047
FRONT END STATIONARY HEAD MODULE							
* STATIONARY HEAD SHELL	1	FE510.1KW	312	312	4,3	1342	1342
* PASS PARTITION	2	FE510.1KW	91	182	4,3	391	783
* STATIONARY HEAD NOZZLE	2	ASTM A105 / A106	44	88	4,1	180	361
* STATIONARY TUBESHEET	1	ASTM A105	2206	2206	4,3	9486	9486
** CHANNEL COVER	1	ASTM A105	1697	1697	4,3	7297	7297
				4485			19268
TUBES MODULE							
* TUBES	1	FE35.2	7327	7327	4,1	30041	30041
* LONGITUDINAL BAFFLE	2	FE510.1KW	760	1520	4,3	3268	6536
* TRANSVERSE BAFFLE	12	FE430	49	588	4,3	211	2528
				9435			39105
				19078			80420

Table 2. Data on the AES architecture

AES ARCHITECTURE	QUANTITY	MATERIAL	WEIGHT [kg]	WEIGHT (TOT) [kg]	EI MAT (UNIT) [mPt]	EI MAT [mPt]	EI MAT (TOT) [mPt]
CENTRAL MODULE							
SHELL	1	FE410.1KW	3614	3614	4,3	15540	15540
SHELL NOZZLE	2	ASTM A105 / A106	44	88	4,1	180	361
SHELL FLANGE REAR HEAD END	1	ASTM A105	664	664	4,1	2722	2722
SHELL FLANGE STAT HEAD END	1	ASTM A105	574	574	4,1	2353	2353
** SUPPORT SADDLE	2	FE410.1KW	101	202	4,3	434	869
				5142			21845
REAR END HEAD MODULE							
** SHELL COVER	1	FE410.1KW	680	680	4,3	2924	2924
** SHELL COVER FLANGE	1	ASTM A105	276	276	4,1	1132	1132
** FLOATING HEAD COVER	1	FE510.1KW	275	275	4,3	1183	1183
** FLOATING HEAD FLANGE	1	ASTM A105	387	387	4,1	1587	1587
** BACKING DEVICE	1	ASTM A105	403	403	4,1	1652	1652
				2021			8477
FRONT END STATIONARY HEAD MODULE							
** STATIONARY HEAD SHELL	1	FE510.1KW	312	312	4,3	1342	1342
** PASS PARTITION	2	FE510.1KW	91	182	4,3	391	783
** STATIONARY HEAD NOZZLE	2	ASTM A105 / A106	44	88	4,1	180	361
** STAT. HEAD FLANGE	1	ASTM A105	561	561	4,1	2300	2300
** STAT. HEAD FLANGE COVER SIDE	1	ASTM A105	528	528	4,1	2165	2165
** CHANNEL COVER	1	ASTM A105	1697	1697	4,3	7297	7297
				3368			14247
TUBES MODULE							
* TUBES	1	FE35.2	8793	8793	4,1	36051	36051
* STATIONARY TUBESHEET	1	FE510.1KW	715	715	4,3	3075	3075
* FLOATING TUBESHEET	1	FE510.1KW	515	515	4,3	2215	2215
* TRANSVERSE BUFFLE	14	FE430	38	532	4,3	163	2288
				10555			43628
				21086			88197

The results of this first type of investigation can be summarised as follows:

- The more complex AES architecture uses more resources and has a greater production impact (overall weight 21086 kg, impact of materials used 88197 mPt) compared to the CFU architecture (weight 19078 kg, impact 80420 mPt).
- Durability and separability of the components allow a first recovery cycle of essentially the same modules for both AES and CFU, but permit a second and final recovery cycle only in the case of AES (in *Tables 1* and *2*, components with a single asterisk are recovered only in the first cycle, those with two asterisks are also included in the second recovery cycle).
- AES architecture gives greater recovery fractions than CFU, as shown by *Graph 1*. The fraction of the first recovery cycle Φ^{EI}_j is greater than 3.4%, despite the fact that the first recovery covers the same main modules in both cases. This evidences a better distribution of the volumes in AES, with an architecture which is more functional for recovery (because, unlike in CFU, the rear end module is not integrated with the central module). The fraction of the second recovery cycle is distinctly greater (+163%), showing a more efficient recovery for AES which, because of its better modularity, allows the second reuse of the entire front end stationary head module.

- As a direct consequence of the better recovery fractions, the Extension of Useful Life values for AES are greater than those of CFU (*Graph 2*), slightly greater after the first recovery cycle (+3.4%) and substantially so after the second and final recovery cycle (+19%).

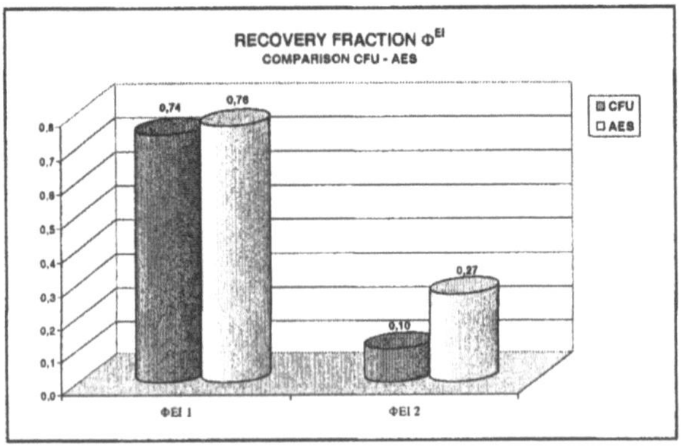

Graph 1. Recovery fraction (CFU-AES)

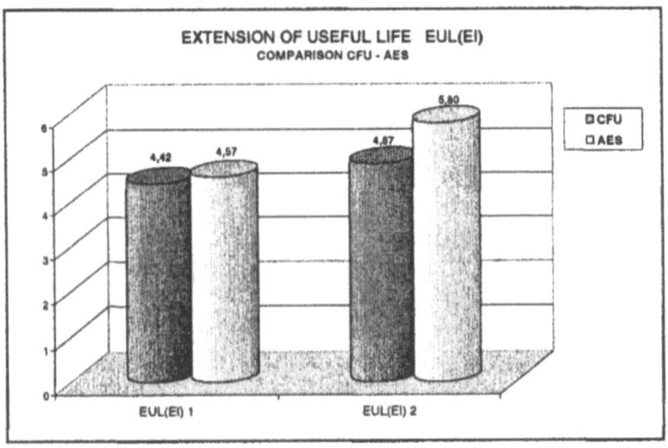

Graph 2. Extension of Useful Life (CFU-AES)

- In conclusion, therefore, the AES construction typology is characterised by a more complex architecture but has better modularity and is more functional for recovery than CFU. This leads on the one hand to the use of more resources and greater production costs, while on the other hand it offers greater suitability for the recovery of these resources.

The second design approach had the aim of evaluating the efficacy of the design tool proposed as an aid to the optimisation of the CFU type architecture. From the comparison with the AES architecture above, it is clear that it would be appropriate to make the tube bundle separable from the front end stationary head module, exactly as in AES, with a reversible flanged coupling which connects the central shell, tube plate and the front end shell. The introduction of this modification is therefore proposed as a design choice which is directed at a greater efficiency of the CFU recovery cycles. This results in the optimised architecture CFU* showed in *Figure 6*, with the principle data reported in *Table 3*.

The results of the comparison between the two architectures, conventional and optimised, can be summarised as follows:

- CFU* is slightly heavier (+0.2%), but also shows a decrease in the overall impact of the materials used (-0.13%) due to a different percentage of materials characterised by different environmental impacts (although the differences are minimal).
- Durability and separability of the components allow the first recovery cycle to cover essentially the same construction modules for both CFU and CFU*, but the second and final recovery cycle is possible only for CFU* (as indicated by the asterisks in *Tables 1* and *3*).
- In the first recovery cycle, CFU* shows a slight decrease in the recovery fraction compared to CFU (-1%), as reported in *Graph 3*. Thus halting at the first recovery operation would mean that the optimisation was inefficient. This slight decrease is however fully compensated in the second recovery cycle, made more effective by a better modularity, with a consequent +92% increase in the recovery fraction of CFU* compared to CFU.

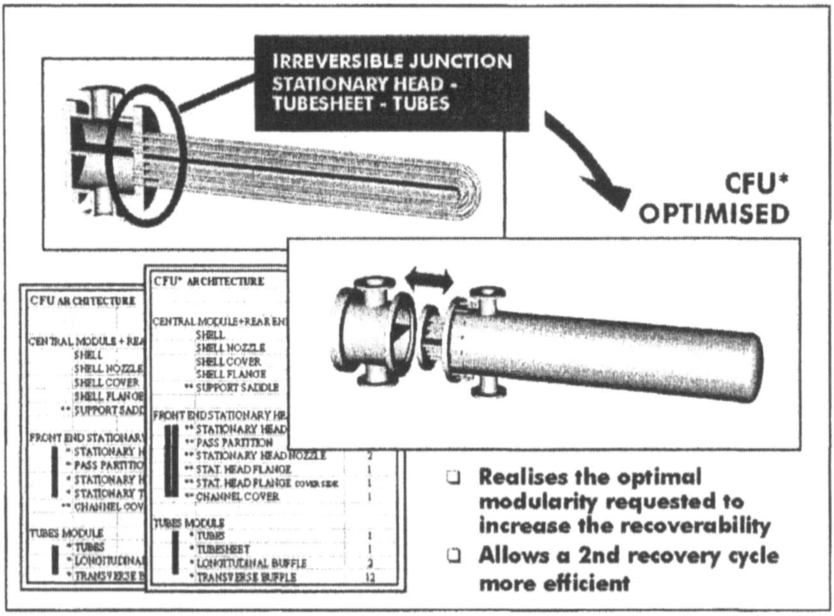

Figure 6. Optimised architecture CFU*

Table 3. Data on the CFU* architecture

CFU* ARCHITECTURE	QUANTITY	MATERIAL	WEIGHT [kg]	WEIGHT (TOT) [kg]	EI MAT (UNIT) [mPt]	EI MAT [mPt]	EI MAT (TOT) [mPt]
CENTRAL MODULE+REAR END HEAD MODULE							
SHELL	1	FE410.1KW	3953	3953	4,3	16998	16998
SHELL NOZZLE	2	ASTM A105 / A106	44	88	4,1	180	361
SHELL COVER	1	FE410.1KW	341	341	4,3	1466	1466
SHELL FLANGE	1	ASTM A105	718	718	4,1	2942	2942
** SUPPORT SADDLE	2	FE410.1KW	101	202	4,3	434	869
				5302			22635
FRONT END STATIONARY HEAD MODULE							
** STATIONARY HEAD SHELL	1	FE510.1KW	312	312	4,3	1342	1342
** PASS PARTITION	2	FE510.1KW	91	182	4,3	391	783
** STATIONARY HEAD NOZZLE	2	ASTM A105 / A106	44	88	4,1	180	361
** STAT. HEAD FLANGE	1	ASTM A105	701	701	4,1	2875	2875
** STAT. HEAD FLANGE COVER SIDE	1	ASTM A105	528	528	4,1	2165	2165
** CHANNEL COVER	1	ASTM A105	1697	1697	4,3	7297	7297
				3508			14822
TUBES MODULE							
* TUBES	1	FE35.2	7327	7327	4,1	30041	30041
* TUBESHEET	1	FE510.1KW	873	873	4,3	3752	3752
* LONGITUDINAL BUFFLE	2	FE510.1KW	760	1520	4,3	3268	6536
* TRANSVERSE BUFFLE	12	FE430	49	588	4,3	211	2528
				10308			42857
				19117			80314

- The results regarding the Extension of Useful Life (reported in *Graph 4*) mirror those of the recovery fractions, therefore again the comparison between CFU* and CFU shows a slight fall in EUL after the first recovery cycle (-1%) and a substantial increase after the second and final recovery (+7.4%). This underlines the importance of using an indicator which extends the evaluation over all the recovery cycles for a complete analysis of the efficiency of the recovery operation.

- In conclusion, the optimisation proposed for the CFU architecture is appreciable only with an evaluation of the entire potentiality of recovery.

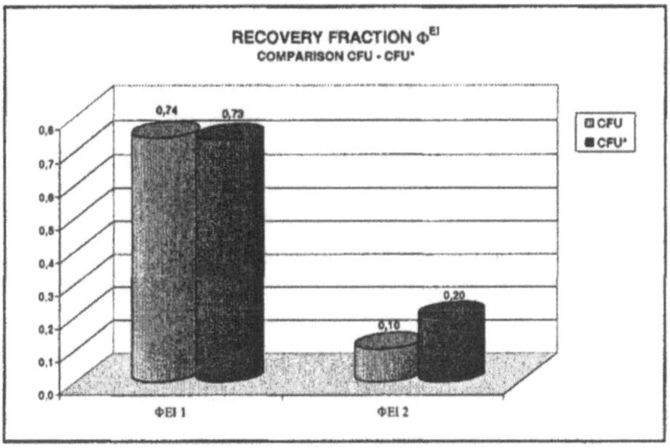

Graph 3. Recovery fraction (CFU-CFU*)

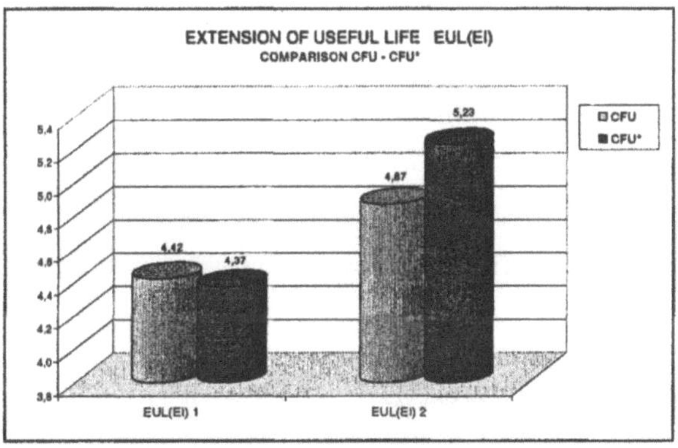

Graph 4. Extension of Useful Life (CFU-CFU*)

Finally, *Graphs 5* and *6* show, respectively, the greater efficiency of the AES architecture compared to the other two examined, in terms of the increases of EUL after the first and second recovery, and the trend of the UL over successive recovery cycles for the three architectures.

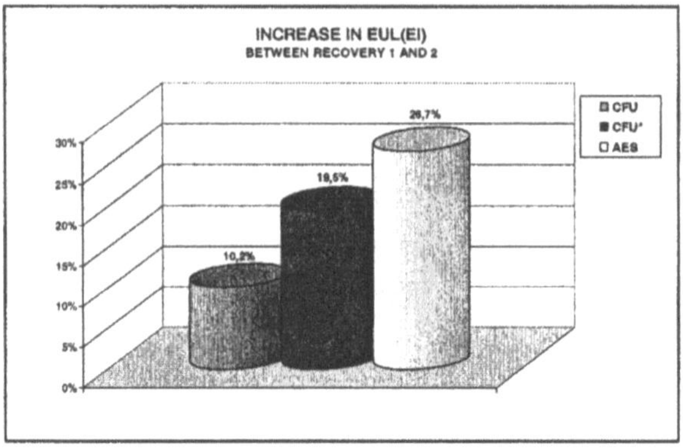

Graph 5. Increase in EUL between 1st and 2nd recovery

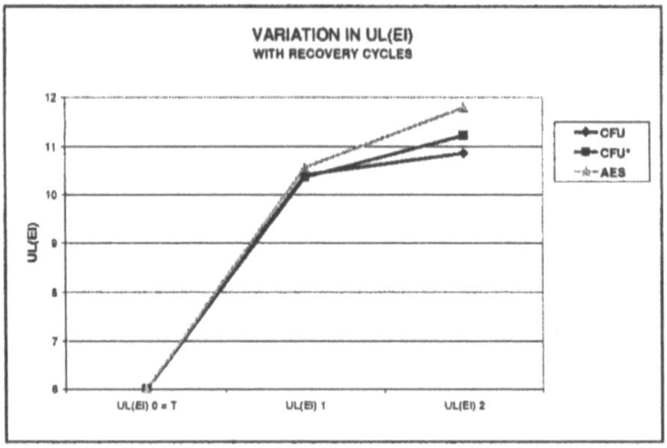

Graph 6. Trend of UL over successive recovery cycles

6. CONCLUSIONS

The study proposes a tool to evaluate design choices directed at the most efficient planning of the recovery cycles for a product at the end of its working life. The methodology developed and the calculation model which sustains it, are based on the evaluation of the determining factors which define the potentiality of each component of a constructional unit according to the possibilities of its recovery and reuse, with the aim of quantifying the benefits of the recovery of resources in terms of the extension of the product's useful life.

Applied as an instrument of analysis and optimisation of the suitability for recovery of the modules characterising the architecture of heat exchangers, the calculation model proposed here seems to confirm the potential for recovery as an aid to the design of recovery cycles, and to suggest its integration with other tools developed for design orientated towards the environmental quality of products.

REFERENCES

Billatos, S., Basaly, N. (1997). *Green Technology and Design for the Environment*, Taylor & Francis, Bristol PA.

Burke, D., Beiter, K., Ishii, K. (1992). Life-Cycle Design for Recyclability, In *ASME Design Theory and Methodology Conference*, Scottsdale.

Dowie, T., Simon, M. (1995). Guidelines for Designing for Disassembly and Recycling, In *R '95 Conference "Recovery, Recycling, Re-integration"*, Geneva.

Giudice, F., La Rosa, G., Risitano, A. (1999). Models and Indicators for the Cost-Benefit Analysis of a Green Product, In *EcoDesign'99 1st International Symposium on Environmentally Conscious Design and Inverse Manufacturing,* Tokyo.

Giudice, F., La Rosa, G., Risitano, A. (1999). Indicators for Environmentally Conscious Product Design, In *EcoDesign'99 1st International Symposium on Environmentally Conscious Design and Inverse Manufacturing,* Tokyo.

Goedkoop, M., Demmers, M., Collignon, M. (1995). *The Eco-indicator 95: Manual for Designers*, Pré Consultants BV, Amersfoort, The Netherlands.

Graedel, T., Allenby, B. (1998). *Design for Environment*, Prentice Hall, Upper Saddle River NJ.

Huang, C. (1996). *Design for X: Concurrent Engineering Imperatives*, Chapman & Hall, London.

Ishii, K. (1995). Life-Cycle Engineering Design, *ASME Journal of Mechanical Design*, vol. 177.

Ishii, K. (1999). Incorporating End-of-Life Strategy in Product Definition, In *EcoDesign '99 1st International Symposium on Environmentally Conscious Design and Inverse Manufacturing*, Tokyo.

Johnson, M., Wang, M. (1998). Economical Evaluation of Disassembly Operations for Recycling, Remanufacturing and Reuse, *International Journal of Production Research*, vol. 36, no. 12.

Kusiak, A. (1992). *Concurrent Engineering: Theory and Practice*, Wiley & Sons, New York.

Lambert, A. (1997). Optimal Disassembly of Complex Products, *International Journal of Production Research*, vol. 35, no. 9.

Navin-Chandra, D. (1991). Design for Environmentability, In *ASME Design Theory and Methodology Conference*, Miami.

Navin-Chandra, D. (1994). The Recovery Problem in Product Design, *Journal of Engineering Design, vol. 5, no.1.*

Tang, Y., Zhou, M. C., Zussman, E., Caudill, R. J. (2000). Disassembly Modelling, Planning, and Application: A Review, In *2000 IEEE International Conference on Robotics and Automation*, San Francisco, CA.

TEMA (1988). *Standards of the Tubular Exchanger Manufacturers Association*, Tubular Exchanger Manufacturers Association, New York.

Index

MIX
Papier aus verantwortungsvollen Quellen
Paper from responsible sources
FSC® C105338

If you have any concerns about our products,
you can contact us on
ProductSafety@springernature.com

In case Publisher is established outside the EU,
the EU authorized representative is:
Springer Nature Customer Service Center GmbH
Europaplatz 3, 69115 Heidelberg, Germany

Printed by Libri Plureos GmbH
in Hamburg, Germany